Space and Society

The Space and Society series explores a broad range of topics in astronomy and the space sciences from the perspectives of the social sciences, humanities, and the arts. As humankind gains an increasingly sophisticated understanding of the structure and evolution of the universe, critical issues arise about the societal implications of this new knowledge. Similarly, as we conduct ever more ambitious missions into space, questions arise about the meaning and significance of our exploration of the solar system and beyond. These and related issues are addressed in books published in this series. Our authors and contributors include scholars from disciplines including but not limited to anthropology, architecture, art, environmental studies, ethics, history, law, literature, philosophy, psychology, religious studies, and sociology. To foster a constructive dialogue between these researchers and the scientists and engineers who seek to understand and explore humankind's cosmic context, the Space and Society series publishes work that is relevant to those engaged in astronomy and the space sciences, while also being of interest to scholars from the author's primary discipline. For example, a book on the anthropology of space exploration in this series benefits individuals and organizations responsible for space missions, while also providing insights of interest to anthropologists. The monographs and edited volumes in the series are academic works that target interdisciplinary professional or scholarly audiences. Space enthusiasts with basic background knowledge will also find works accessible to them.

More information about this series at http://www.springer.com/series/11929

William Sims Bainbridge

Computer Simulations of Space Societies

William Sims Bainbridge
Arlington, VA
USA

ISSN 2199-3882 ISSN 2199-3890 (electronic)
Space and Society
ISBN 978-3-030-08042-6 ISBN 978-3-319-90560-0 (eBook)
https://doi.org/10.1007/978-3-319-90560-0

Cover design: Paul Duffield

Printed on acid-free paper

This Springer imprint is published by the registered company Springer International Publishing AG part of Springer Nature
The registered company address is: Gewerbestrasse 11, 6330 Cham, Switzerland

Contents

Chapter 1
A Virtual Launch into a Computational Cosmos

This book concerns computer simulation of social behavior related to spaceflight, which is a diverse cluster of topics having no simple definition, but potentially very significant implications for the human future. In a 1973 report, NASA said that astronauts had been trained in simulators for a total of 38,261 h for the Mercury, Gemini and Apollo Programs (Woodling et al. 1973). But these involved physical simulators, having electronic instrumentation, but not the totally virtual simulations that will be the main focus of this book, and they did not emphasize the social factors. To be sure, some simulations were the equivalent of dry runs, or rehearsal sessions, in which flight crews performed roles in a scripted division of labor, but much of that activity over the history of space exploration is usually classified as training, not as computer simulation.

1.1 The Evolving Meaning of Simulation

Consider this apparently clear question: What is a computer simulation? Many plausible definitions could be offered, differing in terms of the complexity of the computational phenomenon, and the particular form of programming that produced it. A picture of a planet displayed on the screen of a computer might fit a permissive interpretation of the simulation idea. Indeed, any picture represents something, rather than being the thing itself, and thus is a simulation.

However, the term "computer simulation" is usually applied to a dynamic representation of reality, rather than a static image. Consider a standard equation for physics: $f = ma$. This can be expanded to: force = mass times acceleration. If the force is the thrust from a rocket engine, and we know the momentary mass of the rocket including the fuel it contains, then the formula will predict the momentary acceleration of the rocket, whether in meters per second or feet per second, depending upon the system of units we are using.

W. S. Bainbridge, *Computer Simulations of Space Societies*, Space and Society,
https://doi.org/10.1007/978-3-319-90560-0_1

But f = ma is not yet a simulation, because it is not inherently dynamic. It becomes dynamic as we enter different numbers into the equation. Suppose we also have a formula for the rate at which the particular rocket uses up its fuel. Then we can put both equations into a computer program inside a loop. Every second, the computer will recalculate the mass of the rocket plus its remaining fuel, as the fuel is expended to produce a constant thrust. The equation f = ma now becomes dynamic, and reveals an increasing rate of acceleration, as the same thrust acts upon a declining mass.

In computer science, the word *algorithm* is often used, rather than *equation* or *function*, and other related terms include *procedure*, *routine*, and *program*. Without great precision, these words refer to segments of computer code that accomplish particular goals, which often require assembling together many smaller components. Some of them are identical to algebraic expressions like f = ma, but others take rather different forms, such as this hypothetical example:

If rocket velocity is high and
rocket direction is down and
rocket and ground are at same location then
display a graphic of explosion;

So far, these examples are not social, and some of the most prominent similar examples from the history of digital computing were not very social either. For example, the ENIAC computer dating from 1943–1946 was designed to calculate the trajectories of artillery, doing so in a series of steps that could reasonably be described as simulation (Stern 1981). While not identical to spaceflight, the simulation was quite comparable to modeling the flight of a spacecraft moving through a complex environment. In an earlier publication, I described this process:

> The problem that motivated the U.S. Army to invest in ENIAC was the need for accurate firing tables for aiming artillery during World War II. Many new models of guns were being produced, and working out detailed instructions for hitting targets at various distances empirically by actually shooting the guns repeatedly on test firing ranges was costly in time and money. With data from a few test firings, one can predict a vast number of specific trajectories mathematically, varying such parameters as gun angle and initial shell velocity. The friction of air resistance slows the projectile second by second as it flies, but air resistance depends on such factors are the momentary speed of the projectile and its altitude. Thus, the accuracy of calculations is improved by dividing the trajectory into many short intervals of time and figuring the movement of the projectile in each interval on the basis of the output of the previous intervals and changing parameters (Bainbridge 2004: 221–222).

In the case of ENIAC, the people operating the computer were experts, including its creators, rather than the general public. Several examples from the early chapters of this book will also concern the development of computer simulations as professional research tools. However, given the severe limitations of the current stage the real space program has reached, academic simulation of related social behavior has been rather rare. Two other areas have been rather more productive, educational computer simulations and space-related computer games.

Among the earliest examples of computer simulations open to the general public was an educational "computer" named the Geniac, dating from 1955, that included two space-related programs in its diverse curriculum. I actually owned one of these challenging but primitive systems in 1956, and explored its capabilities for many weeks. Designed by Edmund C. Berkeley, who co-founded the currently influential Association for Computing Machinery back in 1947, it was a modifiable set of rotary switches assembled and wired by the user to work through various problems that could be expressed in formal logic. As the Wikipedia article for Geniac explains, "The name stood for 'Genius Almost-automatic Computer' but suggests a combination of the words *genius* and *ENIAC* (the first fully electronic general-purpose computer)."[1]

The basic structure was a masonite board and six discs, containing holes to which wires and other components could be bolted, notably a battery and several flashlight bulbs. The instruction manual showed how to set up and use 33 "simple electric brain machines," which were circuits that simulated a variety of decision problems. Various numbers of the disks would be bolted to the board, in such a way they could be rotated, representing data input, with one or more lights representing output. Each problem required a particular wiring diagram, such that physically adding bolts and wires was the mode of programming a simulation, always starting from scratch, with no ability to load and unload software as in modern computers. Problem number 8 was "Machine for a Space Ship's Airlock," described thus in the instruction manual:

> The airlock of a space ship has: an inner door that goes from the airlock to the inside of the space ship; an outer door which goes from the airlock to the surface of the strange planet which is assumed to have no atmosphere; a pump which pumps the air from the airlock into the space ship; a valve which allows air from the spaceship to flow into the airlock; and a pressure gage which reports the air pressure in the airlock and may be either high or low. There are four lights in the airlock: safe to open the inner door; safe to open the outer door; dangerous to open either door, conditions OK; dangerous to open either door, conditions bad. We want a warning circuit and automatic locks corresponding (Garfield 1955).

The circuit used three disks, with wiring giving each just two positions: (1) Valve from spaceship to airlock: shut or open, (2) Pump from airlock to spaceship: on or off, and (3) Gage showing pressure in airlock: full pressure or zero. A fundamental assumption was that if it was safe to open one of the airlock doors, the other one would be locked, but both would be locked if it was unsafe to open either. For example, if pressure in the airlock is zero, the value is shut, and the pump is off, then it is safe to open the outer door but not the inner door.

This simulated spaceship apparatus tells one astronaut whether it is safe to open one door or the other, but it is never safe to open both. We could imagine a social dimension, if that astronaut is wearing a spacesuit and is currently in the airlock, while a fellow crew member is inside the ship and not wearing a spacesuit, a complexity the simulation apparently assumes but does not explicitly describe.

The second Geniac space simulation is more social but perhaps less realistic, "The Uranium Shipment and the Space Pirates." Number 23 in the increasingly difficult

[1]en.wikipedia.org/wiki/Geniac, accessed May 2017.

set of 33, it requires five of the six disks, each with just two positions but requiring a rather complex circuit to handle all the combinations:

> A uranium shipment from one of Jupiter's Moons, Callisto, to Earth consists of a freighter rocket ship loaded with uranium and a fighter escort rocket ship disguised as a freighter. Space pirates are known to be lurking on one of the two asteroids, Pallas or Hermes. The pirates suspect that one of the rocket ships is a disguised fighter; therefore they may either attack the first ship or wait in hiding for a second ship. The commander of the uranium shipment can send either ship by the Pallas or the Hermes route and can send the fighter either first or second. If the pirate attacks the fighter, the pirate will be destroyed. If the pirate attacks the uranium ship and the fighter has already passed or taken the other route, then the pirate captures the uranium. If the pirate attacks the uranium ship, and the fighter is taking the same route, and is behind the uranium ship, the pirate is destroyed but during the battle, the pirate destroys the uranium ship. Of course, if the pirates do not attack, there is no combat (Garfield 1955).

There are four outcomes: (1) pirates destroyed, shipment safe, (2) no combat, (3) pirates and shipment both destroyed, and (4) pirates capture the uranium. There are two ways in which this 1955 Geniac simulation foreshadows common episodes in the space-related computer games of five and six decades later. First, the outcome of a social interaction is determined by a complex algorithm. The main difference today is that players cannot themselves predict the outcome with any real confidence, because the algorithm includes a random number, and the rules built into the algorithm are concealed from the player.

Second, the excitement associated with the episode requires a degree of simulated violence, in this case life or death outcomes for the pirates and the crew of the uranium ship. Academic social and computer scientists might prefer to avoid violence, and many would argue that any real combat in outer space would not involve human beings, but be millisecond-brief encounters between robot missiles and information-collecting satellites. As intellectually frustrating as it may be at times, humans require a good deal of emotional arousal to motivate their involvement in computer simulations, and violent combat has become a main feature of the examples considered in the later chapters of this book.

1.2 Computational Social Science

In his recent textbook, *Introduction to Computational Social Science*, Claudio Cioffi-Revilla defines this emerging field as "the interdisciplinary investigation of the social universe on many scales, ranging from individual actors to the largest groupings, through the medium of computation (Cioffi-Revilla 2014)." He maps the field into three subdomains: (1) *Big data*, which refers to the analysis of very large datasets concerning human behavior, requiring advanced computational facilities to manage potentially terabytes of data and very demanding analytical algorithms. (2) *Social networks*, which become both theoretically revealing and methodologically demanding as the number of individuals rises into the thousands and the relationships between them expand into multiple dimensions. (3) *Computer simulation*, which he covers in three distinct chapters, describing methodology, variable-oriented models, and object-oriented models.

Cioffi–Revilla explains that as a methodology, computer simulation of social behavior permits analysis of much greater and thus more realistic complexity than do fixed-form mathematical expressions like f = ma. "This is accomplished by building a computer model of the social system or process under investigation—a virtual world representing relevant aspects of reality—and using that model to perform many kinds of analyses (Cioffi-Revilla 2014)."

One of the first historical examples he cites, Jay Forrester's *World Dynamics*, was my own introduction to this field, in a sociology seminar at Boston University that used pre-publication page proofs of this 1971 book as its text (Forrester 1973). Forrester's Wikipedia page describes him as "the founder of *system dynamics*, which deals with the simulation of interactions between objects in dynamic systems."[2] To give but one example of how Forrester's approach can be applied to spaceflight-related social dynamics, he directly stimulated the highly influential and controversial 1972 study, *The Limits to Growth* (Meadows 1972). It used complex-system computer simulation, anchored in some empirical variables about the resources and economic dynamics of our planet, to develop scenarios of the possible futures for humanity, efficiently summarized on the study's Wikipedia page:

> The original version presented a model based on five variables: world population, industrialisation, pollution, food production and resources depletion. These variables are considered to grow exponentially, while the ability of technology to increase resources availability is only linear. The authors intended to explore the possibility of a sustainable feedback pattern that would be achieved by altering growth trends among the five variables under three scenarios. They noted that their projections for the values of the variables in each scenario were predictions "only in the most limited sense of the word," and were only indications of the system's behavioral tendencies. Two of the scenarios saw "overshoot and collapse" of the global system by the mid to latter part of the 21st century, while a third scenario resulted in a "stabilized world."[3]

The controversies surrounding *The Limits to Growth* are many, most relevant here the possibility that aggressive scientific and technological innovation might expand the scope of the complex human system far beyond the planet Earth, thus becoming limitless. We shall explore simulations of this possibility in later chapters, but we could also consider how Forrester's method can be applied to analysis of the limits to growth not merely on Earth, but on Mars and indeed on any colonized planet. One step further in the development of that particular orientation toward social science computer simulation would consider the possibility that future civilizations might not consist of networks of colonized planets, each limited primarily to its own resources, but exist on moons and in asteroid belts, such that commerce between settlements would be physically easier and thus cheaper than between high-gravity planets.

---

[2]en.wikipedia.org/wiki/Jay_Wright_r, accessed April 2017.

[3]en.wikipedia.org/wiki/The_Limits_to_Growth, accessed December 2017.

The distinction made by Cioffi-Revilla between variable-oriented models, and object-oriented models is logical but not the only way to categorize the alternatives. He refers again to Forrester's work as an influential example of variable-oriented models, but I naturally think of them as an application of multi-variable statistical analysis, such as these traditional techniques of statistical social science: multiple regression, partial correlations, log-linear analysis, and path diagrams. In their more complex forms, these often distinguish independent variables from dependent variables, with relatively independent or dependent variables intervening between them in a complex diagram, somewhat reminiscent of a family tree in genealogy, but one exhibiting considerable incest. These methods become dynamic models if feedback loops are added, or some independent variables are constantly updated by a stream of new information inputs.

The object-oriented model approach is one I have often used, through what are called *multi-agent systems*, in which each agent represents a human being, perhaps possessing artificial intelligence to some degree, interacting with many other semi-autonomous agents. As a practical matter, each agent is typically represented by a structured set of memory registers in the computer, acted upon dynamically by the same set of theory-based algorithms, while the agents interact with each other. One of the most influential examples Cioffi-Revilla described is the cellular automata method of Thomas Schelling, and the widely disseminated *Game of Life* invented by John Conway (Schelling 1960).

Place checkers at random, or in some aesthetically interesting pattern, on a checkerboard. This will be a very simple example of a general principle, and when I programmed an especially ambitious simulation based on this classic idea, it modeled a board with 163,216 squares rather than the traditional 64. The squares on the board are the cells in the cellular part of the name *cellular automata*, and the automatic part is the mechanistic application of a set of behavioral rules to any kind of change in the pattern of checkers. In a demographic simulation, the checkers might represent people, and when two of different kinds are in neighboring squares, the algorithm could add new checkers near them representing their children. Or, each square could be a person, and the presence of a checker on a square could represent an idea in that person's mind, such as a political orientation or religious belief that might be influenced by the adjacent squares representing other people. The original *Game of Life* called the checkers *counters* and abstractly modeled a hypothetical biological system with these three rules:

1. Survivals. Every counter with two or three neighboring counters survives for the next generation.
2. Deaths. Each counter with four or more neighbors dies (is removed) from overpopulation. Every counter with one neighbor or none dies from isolation.
3. Births. Each empty cell adjacent to exactly three neighbors—no more, no fewer—is a birth cell. A counter is placed on it at the next move.

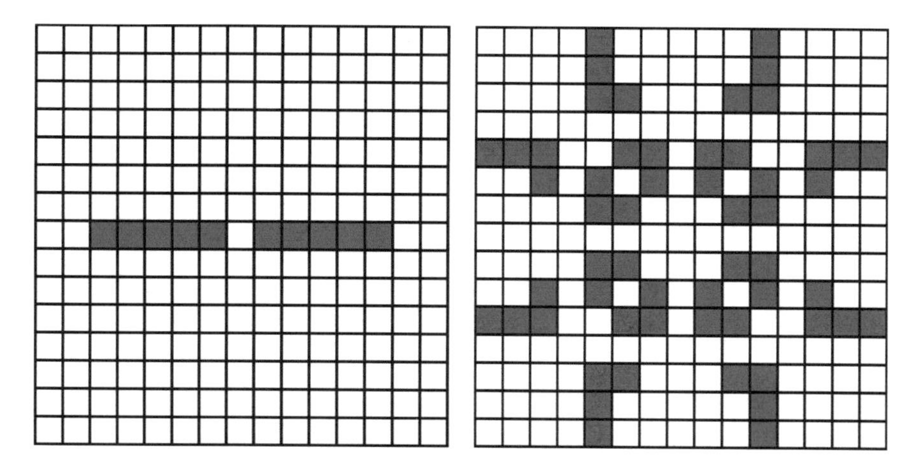

Fig. 1.1 A game of life at step 1 (left) and step 22 (right)

This game does not seem to model life as we know it on Earth, apparently requiring three parents rather than two, so one might claim that it is a simulation of an extraterrestrial life form. Of course, it is very simplistic. Indeed, that is the real value of this example, because depending upon the starting conditions, the configuration may evolve through many steps, leading to what appear to be more complex patterns. This demonstrates how complexity may evolve from simplicity in a rule-based dynamic system. For example, Fig. 1.1 shows a Game of Life that starts with 10 colored squares in a simple pattern on a large checkerboard.

How the pattern of occupied squares evolves depends entirely of course upon the starting situation, yet often with results a human finds surprising. In the case illustrated in Fig. 1.1, starting with the particular 10 squares never leads back to that pattern, nor to total occupancy of all the 225 cells. Rather, the pattern expands to the 56 cells colored in the right-hand configuration, then constantly recycling: 72, 48, 56 that repeats forever. A different arrangement of 10 colored cells leads to a complex scattering of 102 cells after 100 turns, that drops down to a different configuration of 56 cells at 200 turns, then reaches 166 cells at 500 turns (Bainbridge 2006). Obviously, this game has little resemblance to human social interaction, but in general this kind of simulation illustrates how complex systems may behave in ways that defy human prediction, even if they are totally deterministic (Eve et al. 1997).

As Cioffi-Revilla documents, many social science simulations follow either the variable-oriented or object-oriented approach, yet many of those described in this book are more complex, not only combining both of the approaches he emphasizes with each other and with other methods including injection of random events, but significantly incorporating human beings into the system. Very different from the Game of Life, but as its name implies rivalling it, is *Second Life*, a non-game virtual world that includes many socially relevant space-related simulations.

Fig. 1.2 The international spaceflight museum in Second Life

1.3 A Creative Virtual World

Since its launch in 2003, *Second Life* has been the most significant virtual world primarily serving as the computational environment for sharing user-created objects, locales, and computer simulation programs. It was by no means the first example of its species, and much honor belongs to *Active Worlds*, which launched in 1995. The relative success of *Second Life* was a result of synergy between the state of the marketplace and the technology in the first years of the twenty-first century, but much credit also belongs to the team that developed it, Linden Lab in San Francisco.

At no cost, a potential user may register with Linden, download some relatively small software, and create an initial avatar. The first time the user logs in, the avatar is in a starter zone of the equivalent of an entire planet, most of which mimics Earth. After wandering around a while to get accustomed to the system, the user may open the search tool of the interface, set it to seek only places rather than people and groups, and enter a search term like "spaceflight." Among the first hits will be the International Spaceflight Museum, and clicking a "teleport" button will send the avatar there.

Figure 1.2 shows an avatar at a literal high point in the museum, the top of a tall pillar, with many full-sized models of real launch vehicles in the background, spread out over a really wide area. Immediately around the avatar is a simulation of the solar system, like an old-fashioned mechanical orrery that depicts the planets quite large in comparison to their orbits, and the sun rather small. The planets move realistically in their orbits, however, and the avatar has gone into levitate mode to hover in mid air as if he were himself a world. This picture is a screenshot taken March 25, 2017, but the exhibits are several years older.

Below and to the right of the avatar is a sign identifying the orrery as the Solar System Simulator. Clicking the mouse of the user's computer on it links to a page in an online wiki, that reports it was built by Troy McLuhan on some "unknown date prior to 2007 December 31." Using the *Second Life* interface's search tool for "Troy McLuhan" fails to turn up an avatar by that name, evidence that combined with the uncertain date of the orrery suggests that he is no longer active in *Second Life*, although most users currently employ pseudonyms. The wiki reports these technical features of the simulation: "The orbital elements (such as eccentricity) are assumed to be constant at the values they had on January 1, 2000. Interplanetary distances use a conversion factor of 0.7 m per Astronomical Unit (AU). Planetary radii use a conversion factor of 0.0246 mm per km. The Sun is not to scale. If it were to scale, you could place 109 Earths side-by-side inside it."[4]

Some distance behind the planet Jupiter, in this image, stands a replica of the Proton booster, which Wikipedia reports was a family of rockets that launched 412 times over the years 1965–2012. The sign identifying the model links to a similar wiki page to the orrery one, saying it was built by Jamey Sismondi on an unknown date prior to 2007 December 31. Searching that name in the *Second Life* database also fails to turn up a current avatar. Thus, the Solar System Simulator is not only a museum of the history of spaceflight technology, but also an historical archive of the history of virtual worlds, probably catalogued early in 2008. The description of this particular exhibit concentrates much about the social history of spaceflight in a very few words:

> The Proton rocket was originally developed in 1965 by the USSR (the former Soviet Union) as one of that country's first non-ballistic missile designs. The Proton was designed by Vladimir Nikolayevich Chelomei to compete with the larger N-1, designed by Sergey Korolev. Intended for use in the Soviet lunar program, the basic Proton rocket, the Proton-K, was the largest Russian launch vehicle to attain operational status. The Proton is a liquid-fueled design that relies upon toxic, hypergolic fuels. It is a controversial and dangerous combination as several vehicles were lost to failure before the booster became the stable workhorse it now is. Over the years, a variety of booster engine combinations were used, resulting in a number of rocket configurations. Three-stage versions of the Proton-K have been used to launch the Russian elements of the International Space Station (ISS), Zarya and Zvezda, in excess of 20 tons each. The latest version of the Proton lifted a DirecTV broadcast satellite in 2005 but a Proton-M launch in March of 2006 resulted in the failure of the payload to achieve geo-stationary orbit.[5]

Apparently, this text is a decade old, because it does not mention the history of Proton rockets after 2006, and we certainly would have expected a current overview to mention the last flight in 2012. Shifting world history is reflected in the reference to the former Soviet Union, and "the Russian elements of the International Space Station." Interaction of individuals and teams is exemplified by the competition between Chelomei and Korolev. The evolution of mass communication technologies is suggested by the reference to "a DirecTV broadcast satellite." Thus the documentation

[4]ism-writeups.wikidot.com/ism-solarsystemsimulator-en, accessed April 2017.

[5]ism-writeups.wikidot.com/ism-proton-en, accessed April 2017.

Fig. 1.3 Celebration of first contact day in *Second Life*, 2009

associated with a static virtual model of a rocket launch vehicle gives it multiple kinds of social significance.

Social life within *Second Life* has many dimensions, but the dynamic interplay between enduring groups of avatars and special events is especially interesting and easy to illustrate in connection with social simulations concerning spaceflight. Figure 1.3 records the April 2009 celebration of First Contact Day in *Second Life*. As a surprisingly significant wiki named Memory Alpha reports, "First Contact Day was a holiday celebrated to honor both the warp 1 flight of the Phoenix and first open contact between Humans and Vulcans on April 5, 2063 in Bozeman, Montana. On Earth, children were given a day off from school, a fact that Captain Kathryn Janeway remembered was really the only way it was celebrated."[6] So, the anniversary of an historical event was celebrated 54 years before it occurred.

I attended this event myself, in the guise of my original *Second Life* avatar, Interviewer Wilber, who was created November 25, 2006, and described it at some length in my recent book *Star Worlds: Freedom Versus Control in Online Gameworlds* (Bainbridge 2016). It was organized by Mike Calhoun, leader of a *Star Trek* group called United Federation Starfleet. It is not merely a fan club, but a futurist organization with a serious vision based on the philosophy of *Star Trek's* creator, Gene Roddenberry, as Calhoun explains on its website: "United Federation Starfleet, or UFS for short, started as a dream.....a dream to realize and bring Gene Roddenberry's vision to life within the realms of a metaverse community. As Mr. Roddenberry said, 'If man is to survive, he will have learned to take a delight in the essential differ-

[6]memory-alpha.wikia.com/wiki/First_Contact_Day, accessed April 2017.

ences between men and between cultures. He will learn that differences in ideas and attitudes are a delight, part of life's exciting variety, not something to fear."[7]

In Fig. 1.3, two dozen members of United Federation Starfleet can be seen, named in the text placed by the interface over their heads. They listen to a rousing speech by Calhoun, who stands before the full-scale model of the Phoenix, the first human spacecraft capable of exceeding the speed of light, using warp drive. The first flight in 2063 triggered *first contact* with the extraterrestrial Vulcan civilization, because the Vulcans had scrupulously obeyed the Prime Directive, a general rule against influencing and thus openly interacting with any civilization that has not yet itself developed interstellar travel. Thus, the values of the *Star Trek* community imagine a complex interplay of cultures that must be allowed to develop naturally, following whatever principles they themselves decide, so long as they do not harm other civilizations. Human civilization must advance over the coming decades, for example in its conception of gender roles, as illustrated by the Janeway mentioned in a quotation. She does not exist, and yet she has a Wikipedia page:

> Vice Admiral Kathryn Janeway is a fictional character in the *Star Trek* franchise. As the captain of the Starfleet starship USS *Voyager*, she was the lead character on the television series *Star Trek: Voyager*, and later a Starfleet admiral, as seen in the 2002 feature film *Star Trek: Nemesis*. Although other female captains had appeared in previous *Star Trek* episodes and other media, she is, to date, the only one to serve as the central character of a *Star Trek* TV series. She has also appeared in other media including books, movies (notably Nemesis), and video games. In all of her screen appearances, she was played by actress Kate Mulgrew.[8]

Thus, Janeway is a legendary hero—"heroine" having become an obsolete concept—who exists both as an abstract principle of women's progress and as a role played by an actor—"actress" having also become obsolete. Figure 1.4 shows one of my other *Second Life* avatars, Barbara Sims—given that "sims" are simulated locations in *Second Life* lingo—standing on a stairway high up the side of a tall exhibit built for the 2012 science fiction convention held annually in *Second Life*. The building below her, in the center, is her Vulcan Anthropological Museum, containing about a hundred exhibits. Most of then she made herself, but many others were donated by fellow Trekkers. At the time, she was a leader of the Vulcan Council in Second Life, a responsibility she relinquished after the convention.

Each of these annual social gatherings required many people to invest time and money, given that there often have been nearly a hundred exhibits comparable to the Vulcan Anthropological Museum. While visiting *Second Life* is free, there is an extensive internal economy, in which people pay real-world money to Linden Lab to get *Linden dollars*, to rent virtual land, and to upload from their computers each texture or image they use to decorate the surface of a virtual object. The 2017 convention celebrated the 10th anniversary of this huge virtual simulation of possible futures.[9]

---

[7] www.ufstarfleet.org/main/message-from-the-cinc/, accessed April 2017.

[8] en.wikipedia.org/wiki/Kathryn_Janeway, accessed April 2017.

[9] slscificonvention.wordpress.com/2016/03/07/teleport-hub/, accessed April 2017.

Fig. 1.4 Some exhibit halls in the 2012 science fiction convention

Most virtual worlds related to spaceflight are not workspaces where users can create their own objects, locations, and social events, but pre-designed "games," although few resemble traditional definitions of structured forms of play. Many multi-player games have been compared to themeparks, like virtual Disneylands, and until we have considered several examples, formal definitions will remain elusive. The next logical step, after our brief glimpse of the *Star Trek* community in *Second Life*, is to consider a virtual version of its entire popular universe.

1.4 The Final Frontier

To provide contrast with the other examples in this chapter, we can compare a popular virtual galaxy based on very different principles, *Star Trek Online*. I have studied STO extensively over the years through participant observation, publishing three book chapters about it, but here a brief survey will use a wiki as its data source. Very many *Star Trek* related amateur wikis exist at Wikia.com, including one about STO, but it has only 694 pages and seems not to have been updated recently, so we shall use the "official" STO wiki, which claims 50,506 pages (13,499 articles) and fully 305,221 registered users.[10] As the more conventional wiki, Wikipedia, reports, each episode of the original television series began with a proclamation: "Space: the final frontier. These are the voyages of the starship *Enterprise*. Its five-year mission: to

[10]startrekonline.wikia.com/wiki/Star_Trek_Online_Wiki, sto.gamepedia.com/Main_Page, accessed March 2017.

explore strange new worlds, to seek out new life and new civilizations, to boldly go where no man has gone before."[11]

Yet ironically, the original *Star Trek* series broadcast its last episode June 3, 1969, over a month before the July 16, 1969, launch of Apollo 11 that first took humans to the surface of the Moon, and the return from the Moon of the last flight, Apollo 17, was on December 14, 1972. I edited a 2009 issue of the journal *Futures*, titled Space: The Final Frontier, commenting in my introduction, "The saga of space exploration has long wandered in a strange twilight period, arguably either dawn or dusk, in which the goal of human expansion into the galaxy is being neither achieved nor abandoned (Bainbridge 2009)." Since then, the twilight has extended another decade, yet *Star Trek* continues to light the way to the stars.

STO is very much oriented toward stories in the style of the *Star Trek* television series, and does not emphasize free galactic exploration. During interstellar travel, users view the equivalent of a dynamic map, selecting where to send their spaceship, and where they might enter one or another of many limited simulations to undertake one or another kind of mission. The areas of planetary surfaces depicted are quite small, and many missions take place in orbiting space stations. Some story-based missions and also free-form activities involve combat against other spaceships, within limited spheres of space. For our purposes here, that kind of mission offers a good introduction to the way STO conceptualizes space travel in a social context. Therefore, we can start with the wiki page for Space Combat,[12] which is organized in this outline of topics:

HUD = Head-up Display
Basic Tactics
Survival
Dealing Damage
Helm
Specialized Tactics
Alpha Strike
Hit-and-Run
Healing and Support
Debuffs

The term HUD is widely used in computer and videogame parlance, derived from aviation, explained thus on its Wikipedia page: "any transparent display that presents data without requiring users to look away from their usual viewpoints. The origin of the name stems from a pilot being able to view information with the head positioned 'up' and looking forward, instead of angled down looking at lower instruments. A HUD also has the advantage that the pilot's eyes do not need to refocus to view the outside after looking at the optically nearer instruments."[13] Figure 1.5 shows an STO space battle from the perspective of the pilot of the bat-shaped ship just below the center of the image.

---

[11] en.wikipedia.org/wiki/Where_no_man_has_gone_before, accessed August 2017.

[12] sto.gamepedia.com/Space_combat, accessed March 2017.

[13] en.wikipedia.org/wiki/Head-up_display, accessed March 2017.

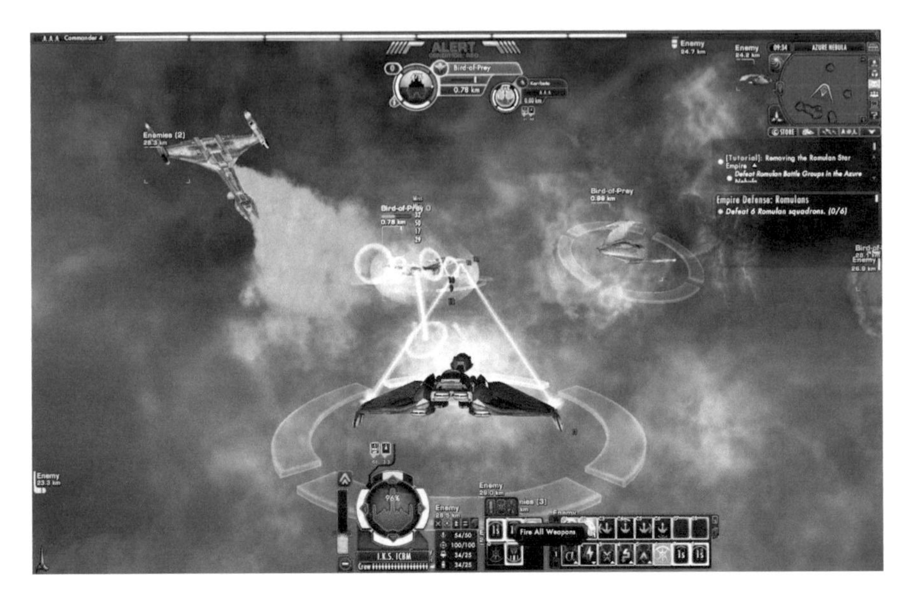

Fig. 1.5 A solo battle in the Azure Nebula

Notice that the player does not see the scene through the eyes of the pilot of the ship, but from a third-person perspective, which distances the player somewhat from the emotional experience but provides information about the situation in a comprehensible form. Also worth noting is the fact that the image presents the scene with a definite up-down dimension. This is nearly universal in virtual depictions of spaceflight, inside the ships as well as outside, even though real space does not establish an up-down distinction. It is wrong to say there is no gravity in outer space, because any large nearby objects place an attractive force upon a spacecraft or human body, but it is not directly perceived as it acts equally upon ship and person. In many cases this lack of realism was caused by the fact that the space-oriented game was programmed inside an already-existing graphics system that was developed to simulate more normal Earth-bound environments. But a second important reason is that the gravity-produced up-down dimension orients human perception, as it was part of the environment in which our species evolved, and most game players would be disoriented by its absence.

The pilot serving as the user's avatar in Fig. 1.5 is a member of the alien but humanoid Klingon species and operating a Klingon ship. A story-related mission sent the pilot into an area of space called the Azure Nebula, which computationally is an *instance* separated from the many other regions depicted in STO. At any given moment there may be many other versions of this instance, functioning in parallel, each experienced by a different player or team of players. This is a training mission, thus rather like a simulation of a simulation, in which the pilot is ordered to defeat battle groups of the hostile Romulan civilization. We see the Klingon pilot engaged in combat with a Romulan Bird of Prey, the two ships firing at each other. Above and to the left an already defeated Romulan ship burns, while above and to the right another one prepares to attack.

The HUD consists of all the diagrammatic shapes arranged around the edges, as well as some tiny words and numbers displayed above the ships. At the top center, where the word ALERT should be visible, is a graphic display of the current status of the combat, notably how much damage the two battling ships have suffered to this point. The upper right corner contains a map and some text stating what the goals of the mission are. The diagrams at the bottom center of the image provide the pilot with extensive information needed to make tactical decisions. The largest segment of this set of diagrams depicts the four active shields that surround the pilot's ship, indicating that the shield in the direction of the enemy has taken damage. As the wiki article explains in its Survival section, the pilot will want to transfer energy to this shield from the others, to prevent this forward shield from failing and allowing serious damage to the ship itself. In this particular battle, the Klingon pilot is not making use of the wiki's Helm advice, which concerns maneuvering, nor of Special Tactics.

Figure 1.6 switches avatars, and now the commander of the primary ship belongs to the United Federation of Planets. He is responding to an emergency message: "Klingon Scouts have been sighted in Pi Canis Sector Space, Captain! We need you to aid us in eliminating this threat. Join in with the concerted efforts of your fellow Starfleet captains in defeating this threat! Go to the Laurentian system, which is located in the Pi Canis Sector Block, and confront a Klingon scout force."[14] The *Star Trek* galaxy is divided into a number of sectors, each 20 light years on a side, and the Laurentian system is in the Xarantine Sector which traditionally was one of three in this Block. The reference to "fellow Starfleet captains" means that this mission must be completed by a team, and Fig. 1.6 shows one in action.

The four circles with graphs along the left side of the image in Fig. 1.6 represent the four other commanders and the current conditions of their ships, allowing each to see who might be in trouble and need assistance. Parts of the display near the lower right corner is a selection of *bridge officer* powers. Bridge officers are *secondary avatars*, that perform very complex social roles interacting with the player's main avatar. In space combat, each of them really functions just as a surrogate for an action the player may select, but in the months before each of them needs to be trained like an actual crewmate to develop the particular power assigned to that crew member.

During quiet time, the player's avatar may inhabit the ship, and then the crew members are found on the command deck, looking like real *Star Trek* characters. For many ground-based missions, players have the option to team up with other players, or use their secondary-avatar crew members. If, for example, a mission requires four avatars on a team, but only two people have decided to perform the mission together, each will bring one computer-simulated crew member along. In ground action, the crew members have a degree of artificial intelligence, but also may be given complex, dynamic sets of instructions by the player.

---

[14]sto.gamepedia.com/Mission:_Slowing_the_Expeditionary_Force, accessed March 2017.

Fig. 1.6 A five-ship team battling Romulans in the Laurentian system

Much of the non-violent action in STO takes place in large space stations called *starbases*. Table 1.1 lists a dozen of them, the first 11 of which were active in early 2017, according to the wiki.[15] Also called *hubs*, they are more than merely places where pilots may dock their spaceships, but also offer ship repair facilities, vendors, markets, trainers, and opportunities for socializing with other players. The first four serve members of all three factions of players, the United Federation of Planets that includes Earth, the Klingons and the Romulans. When STO launched, only the Federation had a complete set of missions, Klingons were a secondary faction, and players could not yet be Romulans. The "first edit" dates in the table report when the first draft of the particular wiki page was posted, thus an indirect measure of how long the starbase has existed in STO. The numbers of authors and edits reflect the fact that wikis are social enterprises. A wiki like this one that reports information derived from a computer simulation like STO is therefore itself a form of social behavior related to spaceflight.

The last starbase in the list, Memory Alpha, offers a remarkable example of the tension between change and stability in human culture. I first visited it, February 6, 2010, only four days after the launch of STO. Wikis tend to preserve their past histories, and the STO wiki page for Memory Alpha was very short when it was created December 3, 2009: "Memory Alpha is the center for learning and crafting

[15] sto.gamepedia.com/Guide:_Starbases, accessed March 2017.

Table 1.1 Twelve starbase hubs in the *Star Trek Online* wiki

Name	Faction	Sector	Authors	Edits	First edit	Last edit
Deep Space 9	All factions	Bajor	35	118	December 3, 2009	October 22, 2016
Delta Quadrant Command	All factions	Delta Quadrant	3	8	November 22, 2014	March 9, 2016
Dyson Sphere Joint Command	All factions	Delta Quadrant	5	9	October 24, 2014	January 14, 2017
Drozana Station	All factions	Donatu	16	55	October 12, 2010	August 28, 2016
Earth Spacedock	Federation	Vulcan	11	103	April 25, 2014	December 3, 2016
Deep Space K-7	Federation	Aldebaran	23	79	January 17, 2010	February 2, 2016
Starbase 39-Sierra	Federation	Sierra	18	68	January 21, 2010	November 5, 2016
Qo'noS: First City	Klingon	Qo'noS	14	86	June 15, 2011	December 26, 2016
Ganalda Space Station	Klingon	Archanis	16	53	February 23, 2010	April 11, 2016
Romulan Flotilla	Romulan	Azure	14	43	May 30, 2013	February 2, 2016
New Romulus Command	Romulan	Azure	8	37	October 15, 2007	June 25, 2013
Memory Alpha	Federation (removed)	Teneebia	24	47	December 3, 2009	July 19, 2014

in Star Trek Online. It will serve as a place to train new Bridge Officers and gain new crafting skills."[16] By March 11, 2011, after major changes in STO that included revision of the crafting system, the page began: "Memory Alpha is the center for crafting for Federation players. The NPCs at Memory Alpha introduce the player to crafting, and enable players to create weapons and equipment by collecting data samples and using one of the research stations provided."[17] The final version of the page, dating from July 19, 2014, begins: "Until its removal from the game on 17 July 2014, Memory Alpha was the center for crafting for Federation players."[18]

The actual birthdate for Memory Alpha was not February 2, 2010, when STO launched, but January 31, 1969, when "The Lights of Zetar," an episode of the

[16] sto.gamepedia.com/index.php?title=Memory_Alpha&oldid=1404, accessed April 2017.
[17] sto.gamepedia.com/index.php?title=Memory_Alpha&oldid=39516, accessed April 2017.
[18] sto.gamepedia.com/index.php?title=Memory_Alpha&oldid=193467, accessed April 2017.

original *Star Trek* television series, was broadcast. The story was set in the year 2269 on the asteroid where the Federation has set up a computer-based library to be the central repository for all its discoveries across the galaxy, called Memory Alpha. It has been attacked by Zetarians, disembodied beings who seize control of the minds of the librarians. The USS Enterprise arrives after the destruction of the library's central computer, whereupon the Zetarians enter the mind of Mira Romaine, a Starfleet science specialist who was on board the Enterprise to deliver equipment to the library. Through her, we learn that the "lights of Zetar" are the few surviving spirits from a dead planet. The fundamental metaphor is that history preserves the "lights" of past generations, which can live again in our own minds. However, our minds are also mortal, and libraries themselves are only temporary.

The *Star Trek* episode was remastered, notably providing a better image of the asteroid from space, for broadcast June 7, 2008. However, by then a real version also existed, in the form of a comprehensive *Star Trek* wiki: "Memory Alpha is a collaborative project to create the most definitive, accurate, and accessible encyclopedia and reference for everything related to Star Trek. The English-language Memory Alpha started in November 2003, and currently consists of 42,358 articles."[19] In a chapter of an earlier book, I suggested the multiple meanings that render this episode a remarkable dramatic work of art:

> "The Lights of Zetar" foreshadowed two features of modern information technology: digital libraries, of which Memory Alpha is an early example, and the use of one "person" by another as an avatar. Of course, spirit possession is an ancient religious idea, and avatars in Hindu religion were the source of the term. But there is another source of this idea. The chief author of the script was Shari Lewis, a puppeteer, ventriloquist, and *Star Trek* fan. In the 1960s, she became famous through very modest but high-quality children's television programs, often centered on a hand puppet of a sheep named Lamb Chop.[20] As her Wikipedia article notes, "Lamb Chop, who was little more than a sock with eyes, served as a sassy alter-ego for Shari."[21] Mira Romaine became an avatar for the Zetarians, but she was also an avatar for Shari Lewis, and an actress named Jan Shutan gave life to the character on the screen.[22] The profusion of overlapping identities in online roleplaying games has a long heritage and is rooted in primary facts of human nature, through which every mentally healthy person is able to play the role of another (Bainbridge 2011).

1.5 Technologically Influential Space Games

Early in 2016, *PCGamer* magazine published brief reviews of "the 50 most important PC games of all time," based on extensive debate among its regular contributors.[23] The criterion for importance was not popularity or financial profits, but the extent to

[19] memory-alpha.wikia.com/wiki/Portal:Main, accessed April 8, 2017.

[20] memory-alpha.org/en/wiki/Shari_Lewis, accessed April 2017.

[21] en.wikipedia.org/wiki/Shari_Lewis, accessed April 2017; the sentence currently reads, "Lamb Chop, which was little more than a sock with eyes, served as a sassy alter ego for Lewis."

[22] memory-alpha.org/en/wiki/Jan_Shutan, accessed April 2017.

[23] www.pcgamer.com/most-important-pc-games, accessed May 2017.

which a game innovated in ways that significantly influenced further developments in the technology or narrative design of entertainment games designed for personal computers. Educational games were generally ignored. For example, the classic series *Reader Rabbit*, dating from 1983, which taught children to read, was not included.[24] Some of the best space-related games were also not included, because they applied to science fiction some technical methods and game design features that had already been developed for fantasy games. For example *Star Wars Galaxies* is famous for giving players the opportunity to build homes on other planets in 2003, but *Ultima Online* had pioneered the building of homes in a fantasy universe already six years earlier. Nonetheless, 10 space-related games were included among the 50, listed here along with a description of the innovation and in parentheses the name of the *PCGamer* author who expressed it:

1962 *Spacewar!* - "Not quite the first computer game ever, but the first to establish that they could exist as more than just Tic-Tac-Toe style programming experiments." (Ed Fries)

1990 *Commander Keen* - "Commander Keen's groundbreaking smooth scrolling hinted at a world where PCs could truly run arcade-style games." (Wes Fenlon)

1992 *Dune II* - "The father of the RTS [Real-Time Strategy] genre." (Tom Senior)

1993 *Frontier: Elite 2* - "Elite gave us the universe. Frontier made it breathe. From heading down to planets (primitive, but still exciting) to filling the spaceways with other pilots, it was like actually entering a whole new universe full of danger, exploration, and profits to be made." (Richard Cobbett)

1994 *Wing Commander III*: Heart of the Tiger - "A space opera elevated by a glittering array of movie talent." (Tom Senior)

1994 *UFO: Enemy Unknown* (X-COM: UFO Defense) - "a simulation rather than a game… a fully contained world…" (Jake Solomon)

1998 *StarCraft: Brood War* - "StarCraft as a whole has fantastic progression, introducing a series of problems and having the player overcome them before moving onto the next problem. Where are the resources? Where am I going to build my base? What's out there in the map? Where's the enemy? How am I going to win?" (Sid Meier)

1998 *Starsiege: Tribes* - "One of the earliest multiplayer-only games… with a pioneering approach to class based action, base defense, vehicles and team-play." (Evan Lahti)

1999 *System Shock 2* - "System Shock refined the shooter/RPG hybrid to a razor's edge. Audio logs, gripping combat, horror and atmospheric storytelling all came together like nothing else." (Chris Avellone)

2003 *EVE Online* - "A whole universe in and of itself, where players, rather than the developers, call the shots. No other MMO has managed to pull it off, from the single-shard where everybody plays, to the depth of tactics and bastardry that goes into high level play." (Phil Savage)

Extremely influential in its day, *Spacewar!* was originally programmed for a PDP-1 commercial computer in 1962 (Graetz 1981).[25] It was innovative in at least three ways: (1) It was a hugely influential example of how an academic computer could be used creatively by "hackers" outside the machine's original design scope. (2) On a primitive video display it simulated spacecraft changing orbits around a star,

[24]en.wikipedia.org/wiki/Reader_Rabbit, accessed May 2017.

[25]www.wheels.org/spacewar/creative/SpacewarOrigin.html, accessed May 2017.

following realistic physics. (3) There were two players, operating two spaceships that could shoot at each other, thus prototyping social simulations. Today, the general public tends to think of hackers as criminals using Internet for nefarious purposes, but originally the term described creative computer programmers, often amateurs or professionals exploring the possibilities outside their work responsibilities.

Several imitations of *Spacewar!* have been produced over the years for a great variety of computer systems, and a replica can be accessed freely online that emulates the original PDP-1 program for any ordinary personal computer. Its description explains: "The code is extremely faithful to the original. There are only two changes. (1)The spaceships have been made bigger and (2) The overall timing has been special cased to deal with varying machine speeds… The 'a', 'd', 's', 'w' keys control one of the spaceships. The 'j', 'l', 'k', 'i' keys control the other. The controls are spin one way, spin the other, thrust, and fire."[26]

The *Commander Keen* series was a commercial children's game, with only limited resemblance to real space travel, yet is honored at the Internet Archive by an emulation that allows anyone to experience this historical case.[27] Wikipedia summarizes the beginning: "In the first episode, 'Marooned on Mars', eight-year-old Billy Blaze, a child genius, builds a spaceship and puts on his older brother's football helmet to become Commander Keen. One night while his parents are out of the house he flies to Mars to explore; while away from the ship the Vorticons steal four vital components and hide them in Martian cities. Keen journeys through Martian cities and outposts to find the components, despite the efforts of Martians and robots; the final component is guarded by a Vorticon."[28] Another form of historical record are walkthroughs of the game, written descriptions of all aspects of playing it including the sequence of required actions.[29]

These first two examples use simple two-dimensional graphics, and emphasize a simulation of the player moving within a virtual extraterrestrial environment, in the form of a spaceship orbiting a star or a boy wearing a football helmet while walking across the Martian surface. The third example, *Dune II*, is a strategy game based on the *Dune* novels by Frank Herbert and set on the Mars-like planet Arakis, where a valuable drug called "spice" can be harvested, but in the context of conflict between feudal factions. As a strategy game it resembles chess to some extent, because the player controls many characters comparable to pawns and makes tactical decisions, rather then being personally represented within the game. It also is available for free online play at the Internet Archive, and has graphics that are more detailed than those of the first episode of Commander Keen, but similar in concept.[30]

---

[26]spacewar.oversigma.com/readme.html, accessed August 2017.

[27]archive.org/details/msdos_Commander_Keen_1_-_Marooned_on_Mars_1990, accessed May 2017.

[28]en.wikipedia.org/wiki/Commander_Keen, accessed May 2017.

[29]www.gamefaqs.com/pc/562691-commander-keen-episode-i-marooned-on-mars/faqs, accessed May 2017.

[30]archive.org/details/msdos_Dune_2_-_The_Building_of_a_Dynasty_1992, accessed May 2017.

Wikipedia reports that *Frontier: Elite 2* "was the first game to feature procedurally generated star systems. These were generated by the game aggregating the mass of material within an early solar system into planets and moons that obey the laws of physics, but which have slightly randomised material distribution in order to ensure each system's uniqueness."[31] Procedural generation requires some explanation. When *Commander Keen* simulated Mars, and *Dune II* simulated Arakis, every detail of the action and every image of the environment had been produced in advance. In procedural generation, pioneered by *Frontier: Elite 2*, each planet is created by algorithms, often guided by input from random number generators, when the user runs the program.

A simplistic example is the color of the planet's surface, which would be predefined as rusty red for Mars and perhaps a more yellow sandy color for Arakis. In procedural generation, each of the display's main colors—red, blue and green—could be represented by a number from 0 to 255. An algorithm could take as input the ID number of a solar system, and the randomly determined distance of a planet from its star, and generate three integers in the 0–255 range that would define the color of the particular planet. Other algorithms could determine the locations of valuable minerals and the characteristics of animal species on the planet. The most recent game in the long-lived *Elite* series, *Elite Dangerous*, launched in 2014, claims to simulate 400 billion solar systems in this way. Chapter 6 of this book will closely examine a controversial online game called *No Man's Sky* that uses procedural generation to simulate a reported 18 quintillion solar systems.

The term *space opera*, used to describe *Wing Commander III* in the quotation above, refers to science fiction that emphasizes action and character archetypes over real science, and thus intermediate between classic science fiction and magical fantasy (Bainbridge 1986). The most popular example is the *Star Wars* saga, and indeed the actor who played Luke Skywalker in that series provided the voice acting for the hero of this computer game. As Wikipedia explains, Wing Commander III "uses extensive live action full motion video to add an interactive movie-style presentation to the space combat gameplay."[32]

Both *UFO: Enemy Unknown* and *StarCraft: Brood War* are strategy games, but UFO is less interesting because it takes place entirely on the Earth, defending it against an invasion by extraterrestrials. *StarCraft*, as I described it in an earlier book, "imagines a distant future in which humanity has spread across many solar systems, gone through a series of interstellar wars, and has come to be dominated by a military dictatorship (Bainbridge 2011)." *StarCraft*, like a number of other very popular games, had a multimodal dimension, as novels were published that expanded upon the stories and their cultural contexts. *Starsiege: Tribes* is a violent, first-person shooter, innovative for its squad-based online multiplayer aspect. Apparently, various versions of it can be found online today, but only from "unofficial" sources.[33]

---

[31] en.wikipedia.org/wiki/Frontier:_Elite_II, accessed May 2017.

[32] en.wikipedia.org/wiki/Wing_Commander_III:_Heart_of_the_Tiger, accessed December 2017.

[33] en.wikipedia.org/wiki/Starsiege:_Tribes, accessed May 2017.

System Shock 2 takes place on the first faster-than-light interstellar ship, named the von Braun after rocket pioneer Wernher von Braun. As its Wikipedia article reports, "The player assumes the role of a soldier trying to stem the outbreak of a genetic infection that has devastated the ship. Like *System Shock*, gameplay consists of first-person combat and exploration. It also incorporates role-playing system elements, in which the player can develop skills and traits, such as hacking and psionic abilities."[34] However, watching a series of YouTube videos in which a player works his way through the game, it appears unrelated to major questions about interstellar travel, being a scavenger hunt for high-tech resources in a maze of rooms and corridors infested by zombies.[35] Earlier I had studied a successor game, titled *BioShock*, which is similar in gameplay but set in an undersea utopian city similarly infested with monsters (Bainbridge 2011).

The last game on the list, *EVE Online*, will be featured in Chap. 6 of this book, and is remarkable for its complexity and the ability of its small team of developers to build it every bigger and better, over now the past 14 years. Also remarkable is the fact that it is the only space game on the *PCGamer* list that launched in the current century. *EVE* is a massively multiplayer online role-playing game (MMORPG) that deserves to be considered a virtual galaxy. When game blogger Justin Olivetti identified his 10 best space games, he listed it first, and offered this general observation: "We yearn to slip the surly bonds of the world to explore the cosmos in our very own rocket ship to see what is out there... It's well-known that sci-fi MMORPGs are in the minority, and only a fraction of those center around or contain some element of space flight and combat. However, over the years we've seen online games here and there allow us to live out our fantasies of being a space jockey, whether in the form of a trader, a fighter pilot, or an explorer (Olivetti 2017)."[36] Historians of space will recognize that the phrase "slip the surly bonds" was poetry quoted by Ronald Reagan when he went on television to honor the astronauts killed in the real-world Challenger disaster.

1.6 A Real World Colonization Simulation

At great effort, it is possible to construct real-world simulations of planetary colonization, with human beings interacting within a physical environment, rather like the training simulations for astronauts but more ambitious. In 1985 I was privileged to work with the widely admired British sociologist, Michael Young, as he assembled a group of volunteers with the hope of creating a year-long simulation of a colony on Mars, called the Argo Venture (Lunan 2002). He had done very influential research on family and community structures, and thus was well prepared to design a simulation that would gain serious scientific knowledge about the possible social evolution of a Mars colony (Young and Willmott 1957). He was also famous for activism,

[34]en.wikipedia.org/wiki/System_Shock_2, accessed May 2017.

[35]www.youtube.com/watch?v=RsnG47JICT4, accessed May 2017.

[36]massivelyop.com/2017/02/23/perfect-ten-mmorpgs-that-help-you-get-your-spaceship-on.

for example playing a key role in the founding of the Open University and writing an influential satire about intellectual inequality titled *The Rise of the Meritocracy*, earning for himself the ironic but genuine title, Baron Young of Dartington (Young 1958).[37]

I interacted extensively with Baron Young while he was visiting Harvard University, then collaborated in a gathering in Britain to plan Argo Venture. He expressed two sociological observations about space exploration: (1) Many nations had been energetically developing their own space technology. (2) Nations could be brought closer together through cooperative efforts in space. Thus, Argo Venture was intended to be a crucial step in the unification of space exploration, toward establishment of a real colony on Mars. In 2010, 8 years after his father's death, Toby Young reported:

> One of the reasons he wanted to do this is because he disapproved of the militarisation of space. This was the era of the Strategic Defence Initiative and Michael wanted to recapture space on behalf of liberal humanitarianism. In his imagination, the colony that would be established on Mars would be peopled by left-wing intellectuals who, after a long discussion about the rights of man, would sign a declaration of independence. The planet would then become a beacon of hope for free thinkers around the world in much the same way that America was and continues to be.[38]

Further insight into the intellectual background of Argo Venture can be read into the fact that James Lovelock was an honored guest at the organizational meeting. In collaboration with several other visionaries, he had developed the Gaia Hypothesis, which Wikipedia says "proposes that organisms interact with their inorganic surroundings on Earth to form a synergistic self-regulating, complex system that helps to maintain and perpetuate the conditions for life on the planet. Topics of interest include how the biosphere and the evolution of life forms affect the stability of global temperature, ocean salinity, oxygen in the atmosphere, the maintenance of a hydrosphere of liquid water and other environmental variables that affect the habitability of Earth."[39]

At the time of the Argo Venture meeting, Lovelock had just published a book titled *The Greening of Mars*, that proposed terraforming the planet's atmosphere by using and thus eliminating the world's supply of ballistic missiles to deliver greenhouse gasses that could increase Mars's surface temperature and thus perhaps also its atmospheric pressure (Lovelock and Allaby 1984). Later writers employed mathematical models to assess this possibility, but for Michael Young it enhanced his social vision for the red planet (McKay 1991). Contributing even more actively to Argo Venture was John Percival, who had already successfully completed a real-world simulation of living under Neolithic conditions for the British Broadcasting Corporation (Percival (1980).

Prior to the meeting, Young had publicized the idea of a simulated Martian colony here on Earth, and encouraged volunteers to apply to join this year-long adventure.

---

[37] en.wikipedia.org/wiki/Michael_Young,_Baron_Young_of_Dartington, accessed May 2017.

[38] www.spectator.co.uk/2010/03/my-father-would-be-pleased-about-the-launch-of-a-british-space-agency, accessed May 2017.

[39] en.wikipedia.org/wiki/Gaia_hypothesis, accessed May 2017.

I prepared a questionnaire to be sent to those who inquired, and our cover letter included a paragraph specifically calling it a simulation:

> This questionnaire will count as an application form for those of you especially interested in joining the simulation as a participant, but please fill it out even if you do not want to participate, because we need your thoughts on many aspects of our project. The information we gain through this questionnaire will not only guide planning for the simulation but may also result in a scientific report helpful to others working along similar lines. Your reply will, of course, be treated in confidence, and your name will not be used in any report without seeking your express approval.

As fate would unfortunately determine, the simulation was never launched, partly because negotiations to use particular British facilities as the location fell through, and because the Biosphere 2 project in Arizona stole much of Argo Venture's thunder, despite the fact that it was more oriented toward preservation of Earth's ecology with only limited ability to simulate a Mars colony.[40] For me, the questionnaire served as a pilot study for the much larger survey project I did in 1986, leading to the book *Goals in Space*, and I used a little of the data to illustrate methods for studying human values in a textbook titled *Survey Research: A Computer-Assisted Introduction* I published in 1989 (Bainbridge 1991).

The concept of *values* is admittedly somewhat controversial in contemporary social science, because it assumes that the behavior of individuals and social groups is shaped by fundamental principles, rather than being the aggregate of many small memories of actions that involved practical costs and benefits (Bainbridge 1994). Yet when asked to express their values, most questionnaire respondents are happy to oblige. Included in the Argo Venture questionnaire was a battery of 18 items developed by Milton Rokeach that asked respondents to rank values (Rokeach 1974). Table 1.2 lists them in the order of mean ranking by Argo Venture respondents, in comparison with summaries of data from a study of commune members by Benjamin Zablocki and some of Rokeach's own reported results.

A total of 133 volunteers and others supportive of Argo Venture submitted application questionnaires, and 122 of them correctly responded to the 18 value items, answering all and never giving two the same ranking. In his book *Alienation and Charisma*, Zablocki had published the aggregate responses from 174 members of small communes (Zalblocki 1980). In his early research on American values, Milton Rokeach had obtained responses in 1968 from 665 men and 744 women. Partly for clarity, and partly because the studies used different formulas to calculate averages, the table compares the collective ranking of the 18 values within each of the four groups of respondents. Ideally, one would want to assemble a dataset of the complete raw data from many studies, and use far more sophisticated methods of analysis.

Given our limited use of the data here for illustrative purposes, we can look at the statistical correlations between the columns of aggregated rankings. The Argo Venture enthusiasts have values rather similar to those of the commune members, as reflected in a correlation of 0.83. Coincidentally, the American men and women have the same high correlation with each other, 0.83. The commune members are very

[40]en.wikipedia.org/wiki/Biosphere_2, accessed May 2017.

Table 1.2 Values ranked by volunteers for the argo simulated mars colony

Value	Argo member mean	Ranking of the values (1–18) by:			
		Argo members	Commune members	Rokeach males	Rokeach females
Wisdom (a mature understanding of life)	6.18	1	2	8	7
Freedom (independence, free choice)	6.42	2	7	3	3
Self-respect (self-esteem)	7.07	3	5	6	6
Inner harmony (freedom from inner conflict)	7.14	4	1	13	12
A world at peace (free of war and conflict)	7.16	5	9	1	1
True friendship (close companionship)	7.26	6	4	11	9
Happiness (contentedness)	7.51	7	6	5	5
An exciting life (a stimulating, active life)	8.01	8	10	18	18
A sense of accomplishment (lasting contribution)	8.37	9	11	7	10
Equality (brotherhood, equal opportunity for all)	8.62	10	8	9	8
Family security (taking care of loved ones)	9.19	11	13	2	2
Mature love (sexual and spiritual intimacy)	9.25	12	3	14	14
A world of beauty (beauty of nature and the arts)	9.76	13	12	15	15
Pleasure (an enjoyable, leisurely life)	13.00	14	15	17	16
Social recognition (respect, admiration)	13.21	15	16	16	17
A comfortable life (a prosperous life)	13.62	16	14	4	13
National security (protection from attack)	14.26	17	18	10	11
Salvation (saved, eternal life)	14.84	18	17	12	4

different from the general American public, having a correlation of only 0.14 with the men and 0.18 with the women. The Argo Venture respondents are less extreme, having a value correlation of 0.36 with American men and 0.40 with American women. Thus, if we simplistically conceptualize the results along a line of deviation from standard norms, the commune members are at the far end away from ordinary citizens, and the Argo Venture people are near them, but in the direction of the conventional respondents.

Certain values are rated very differently, in some cases challenging theories of human values we might have. The very last thing Americans seemed to want in 1968 was an exciting life, while the two radical groups rated it much higher, if not near the top. Americans place family security in second place, while the Argo and commune respondents place it just below the middle of their rankings. There were also differences on wisdom, inner harmony, and a world at peace, that might suggest the Argo and commune respondents valued personal harmony with world harmony. We might summarize these comparisons by suggesting that members of radical social experiments favor personal adventure in ways the average person does not, and devalue conventional family life. As reasonable as this conclusion may seem, we must be aware that the Rokeach approach to measuring values is only one of many, useful here as an example of how real people involved with social experiments could be studied scientifically.

1.7 Conclusion

This chapter provides a preparation for those to follow, by offering an overview of the range of space-related simulations worthy of study. The next two offer historical and technical backgrounds, Chap. 2 by focusing on small-scale simulations of individuals interacting in the process of colonizing Mars, and Chap. 3 by looking more deeply at the question of how cultural lag may influence interstellar colonization or communication. Chapter 4 surveys some of the recent and current educational simulations that have relevance for the evolution of an interplanetary civilization. Chapter 5 goes beyond the limits of our solar system through strategic games that simulate the kind of thinking required to go where no one has gone before. The four concluding chapters explore the vast cultural territory of massively multiplayer online space games, including *EVE Online*, *No Man's Sky*, *Tabula Rasa*, *Star Wars: The Old Republic*, *Anarchy Online*, *Entropia Universe*, and *Mass Effect: Andromeda*. Each of these is a compromise between scientific accuracy and human hopes, but has great intellectual depth that can help us better understand the real human future.

What valuable goals can be achieved by computer simulation of social activities related to spaceflight? In some cases, but currently not many, a simulation may contribute to the design of a technological system. That was the goal of ENIAC, developing a system of accurate artillery for efficient warfare. Arguably, *The Limits to Growth* could have been the first step in designing a sustainable society, yet so far that has not been achieved. Also, the technological infrastructure for interplanetary colonization, which is the precondition for most of the simulations reported in this book, does not exist yet, and even its fundamental design principles remain unknown. Will colonization of Mars somehow be so profitable economically, or human transcendent motives so powerful, that the extreme cost of exceedingly inefficient chemical rockets will cease to be prohibitive? Or will some old dream such as nuclear rockets be fulfilled, reducing the transport cost such that expanding into the solar system will become affordable? Today, we can simulate extraterrestrial colonies in the abstract, but lacking a clear plan to render them viable in the real world, we cannot be very precise in programming realistic simulations.

With occasional exceptions in economics and demography, computer simulations in social science tend to be theoretical explorations, helping us think through general issues and identifying implications of abstract assumptions, but rather detached from practical applications. Yet in both education and entertainment, space simulations have been rather popular. Several kinds of space simulation can teach the facts of astronomy, and including humans with meaningful stories can enhance their educational power with many students. Yet the cognitive benefits go beyond astronomy, to include logical planning, observation of the features of unfamiliar environments, and operating complex computer software. Commercial computer games constitute a major new artform, structured as simulations but perhaps better named *stimulations*, for the aesthetic and emotional benefits they generate. Beyond all these immediate human meanings, many examples reported in this book provide a subjective bridge between today's user and the distant human future, including the possibility that when interstellar travel is finally achieved centuries hence, it will be our minds but not our bodies that explore the distant stars.

References

Bainbridge, William Sims. 2006. Transformative Concepts in Scientific Convergence. In *Progress in Convergence: Technologies for Human Wellbeing*, ed. William Sims Bainbridge, and Mihail C. Roco, 24–45. New York: Academy of Sciences.

Bainbridge, William Sims. 1994. Values. In The Encyclopedia of Language and Linguistics ed. R.E. Asher and J.M. Y. Simpson, 4888–4892. Oxford, Pergamon.

Bainbridge, William Sims. 2004. ENIAC, in *Berkshire Encyclopedia of Human-Computer Interaction*, ed. William Sims Bainbridge, 221–222. Great Barrington, Massachusetts: Berkshire.

Cioffi-Revilla, Claudio. 2014a. *Introduction to Computational Social Science*, 2. London: Springer.

Cioffi-Revilla, Claudio. 2014b. *Introduction to Computational Social Science*, 225. London: Springer.

Eve, Raymond A., Sara Horsfall, and Mary E. Lee (eds.). 1997. *Chaos and Complexity in Sociology: Myths, Models and Theory*. Thousand Oaks, California: Sage Publications.

Garfield, Oliver. 1955a. *Geniacs: Simple Electric Brain Machines, and How to Build Them*, 12. New Haven, Connecticut: Oliver Garfield Company.

Garfield, Oliver. 1955b. *Geniacs: Simple Electric Brain Machines, and How to Build Them*, 30. New Haven, Connecticut: Oliver Garfield Company.

Graetz, J.M. 1981. The Origin of Spacewar. *Creative Computing*. 6 (8): 56–67

Jay, W. 1973. *Forrester, World Dynamics*. Cambridge, Massachusetts: Wright-Allen Press.

Lovelock, James, and Michael Allaby. 1984. *The Greening of Mars*. New York: Warner Books.

Lunan, Duncan. 2002. Lord Young of Dartington and the Argo Venture. *Space Policy* 18: 163–165.
McKay, Christopher P., Owen B. Toon, and James F. Kasting. 1991. Making Mars Habitable. *Nature* 352: 489–496.
Meadows, Donella H., and Dennis L. Meadows. 1972. *Jørgen Randers and William W*. Behrens III, The Limits to Growth. New York: Universe Books.
Olivetti, Justin. 2017. Perfect Ten: MMORPGs that Help You Get Your Spaceship On," Massively Overpowered, February 23, 2017.
Percival, John. 1980. *Living in the Past*. London: British Broadcasting Corporation.
Rokeach, Milton. 1974. "Change and Satability in American Value Systems, 968–1971. *The Public Opinion Quarterly* 38 (2): 222–238.
Schelling, Thomas. 1960. *The Strategy of Conflict.* Cambridge, Massachusetts: Harvard University Press; *Micromotives and Macrobehavior* (New York: Norton, 1978); Martin Gardner. 1970. The Fantastic Combinations of John Conway's New Solitaire Game "Life". *Scientific American* 223:120.
Stern, Nancy, B. 1999. *From ENIAC to UNIVAC: An Appraisal of the Eckert-Mauchly Computers (Bedford, Massachusetts: Digital Press, 1981); Scott McCartney, ENIAC, the Triumphs and Tragedies of the World's First Computer*. New York: Walker.
William Sims, Bainbridge. 2016. *Star Worlds: Freedom Versus Control in Online Gameworlds*, 213–214. Ann Arbor, Michigan: University of Michigan Press.
Bainbridge, William Sims. 2009. Space: The Final Frontier. *Futures* 41: 511–513.
Bainbridge, William Sims. 2011a. *The Virtual Future*, 97. London: Springer.
Bainbridge, William Sims. 1986. *Dimensions of Science Fiction*, 124–129. Cambridge, Massachusetts: Harvard University Press.
Bainbridge, William Sims. 2011b. *The Virtual Future*, 9. London: Springer.
Bainbridge, William Sims. 2011c. *The Virtual Future*, 7–9. London: Springer.
Bainbridge, William Sims. 1989. *Goals in Space: American Values and the Future of Technology (Albany, New York: State University of New York Press, 1991), Survey Research: A Computer-Assisted Introduction*. Belmont, California: Wadsworth.
Woodling, C.H., Stanley Faber, John J. Van Bockel, Charles C. Olasky, Wayne K. Williams, John L. C. Mire, and James R. Homer. 1973. *Apollo Experience Report: Simulation of Manned Space Flight for Crew Training*, 4. Washington, DC: NASA.
Young, Michael, and Peter Willmott. 1973. *Family and Kinship in East London (London: Routledge and Kegan Paul, 1957); The Symmetrical Family*. London: Routledge and Kegan Paul.
Young, Michael. 1958. The Rise of the Meritocracy, 1870–2033. London: Thames and Hudson.
Zalblocki, Benjamin. 1980. *Alienation and Charisma*. New York: Free Press.

Chapter 2
Simulated Martian Social Science Laboratories

Academic computer simulation of space-related social behavior dates back at least three decades, and has evolved in very interesting ways over its lifetime. Therefore this chapter and the next will provide the background for understanding today's developments by examining first some models of interplanetary colonization, then some possible parameters of interstellar colonization and communication. In its early days, computer simulation of social behavior was primary considered a means for rendering formal theory more logically rigorous, without any assumption that computers could model and therefore predict the dynamics of actual societies. Today, there is much room to debate whether computer simulation has evolved sufficiently, in the context of vastly more powerful computers, or is still limited by the lack of sufficiently good abstract theories and adequate data about the real world. Given those doubts, this book will use this chapter and the next as preparation for exploring virtual worlds that are only partially computerized, because they also include real human beings and are shaped by the humanity of both their designers and their users.

2.1 A Sociological Laboratory

To provide conceptual and historical background, this chapter expands upon a series of admittedly primitive simulations written by the author for *Sociology Laboratory*, a pioneering 1987 textbook on social theory based in computer simulations that were included with the book (Bainbridge 1987). One simulation asked students to imagine that two NASA-supported labs, JPL and Ames, were competing to see which could achieve a designated innovation first, modeling the behavior of small teams of researchers. Another employed a computational method inspired by sociobiology to imagine the behavioral evolution of a hypothetical extraterrestrial life-form inhabiting Neptune's largest moon. Three of the most advanced simulations imagined building a society from scratch. So, to add plausibility as well as interest for students, they were presented as the first expeditions to Phobos and Deimos, followed by establishment of a very small colony on Mars. Another simulation modeled competing

W. S. Bainbridge, *Computer Simulations of Space Societies*, Space and Society,
https://doi.org/10.1007/978-3-319-90560-0_2

subcultures in the context of sociological theories of deviant behavior, thus presented as happening in the terrestrial city where many of the theorists had worked, but just as easily conceptualized as a reasonably well-established Mars colony. They illustrated general principles of human cooperation, division of labor, and the emergence of cultural models of reality that not only supported efficient economic exchange, but even encouraged development of religious beliefs. A recent article published in *Space Policy* asserted "we cannot predict the cultural evolution of Martian politics, society, and its legal system," but simulation can certainly be used to explore alternative possibilities (Szocik et al. 2016).

The textbook began with a rhetorical point that may even be more meaningful three decades later. Until some time around 1980, sociology professors and other academics teaching social science classes habitually sent their students out of the classroom to practice observational and even experimental research in the real world, often enlisting inexperienced undergraduates as field researchers. While various practices exist today, there is a much greater concern about the ethical issues concerning the privacy of people being observed, and a host of other practical challenges. By 1987 it was quite clear that many kinds of sociological research were either unethical or impractical for undergraduate assignments, so computer simulations seemed an alternative worthy of exploration.

Quite apart from their educational value, computer simulations are an especially valuable way of developing and evaluating the logical structure and implications of social theories. Naively, the first question people tend to ask about a theory concerns whether it is true - whether it matches the facts observed in the real world. Yet an equally important question that may even be more primary concerns whether the theory is logical - whether its components are really integrated into an intellectual system, especially one capable of producing insightful new ideas based on deduction from the ideas out of which it was constructed.

At a first approximation, social theories may be either formal or interpretive. A formal theory is an integrated set of abstract rules describing how a system of real phenomena operates. An interpretive theory expresses the meanings humans attach to direct experiences, stated in ordinary language and through metaphors. Each may provide insights that contribute to the development of a theory belonging to the other category, but by nature they are very different. Computer simulation harmonizes better with formal theory, although many of the examples later in this book also simulate interpretive theory.

The seventh simulation in *Sociology Laboratory*, called Invent, was designed to give students a simple introduction to the ultimately very complex interaction between social structure and cultural structure, imagining Jet Propulsion Laboratory as the location for the action (Koppes 1982). JPL is the historical center for much of the exploration of the solar system by robot probes, and I did observational field research there in 1981 and 1986 during the encounters of the Voyager 2 space probe with Saturn and Uranus (Bainbridge 1991). Realistically, we can imagine that JPL is a dynamic collection of small research and development teams, each of which takes on a series of special assignments.

Each discovery or invention is a combination of smaller ideas that fit together to make the final result. These are *cultural elements,* concepts produced by many different people and communicated through social exchanges to other people. Thus, the person who first works out a new discovery or invention is not just somebody with a powerful imagination but somebody who has received many relevant ideas via communication with other persons.

The Invent simulation imagined there are five to nine members in the typical team. Each of them works partly alone and partly as a member of the group. Each tries to find particular pieces of the puzzle and test these ideas in research. When they have found good ideas, they share them with the particular other members of the team with whom they have close communication ties. Those people will pass the ideas on until all members have them, but that process of information diffusion takes time. Once an individual has several of the pieces of the puzzle, whether personally discovered or received via communication, he or she begins trying to fit them together. Eventually, somebody on the team will put the last piece in place, and JPL will have its answer. Some individuals are good at inventing, whereas others can merely communicate ideas from one person to another, but communicating is a very important function in the discovery business.

Given the goal of using simple computer simulations to instruct undergraduate students in the principles of social theory, the Invent simulation was naturally very simple. Instead of really making astronomical discoveries or devising new spacecraft technologies, the simulated JPL team members merely guessed words, one letter at a time, assembled into growing combinations of letters.

Another reason for simplicity was the available technology at the time. The software was programed for IBM personal computers, and compatibles by other manufacturers, with 128 K of memory and using the DOS 2.0 operating system. Many of the IBM computers in colleges at that time lacked real graphics capabilities, so the display had to mimic graphics by moving ordinary characters around on the screen. While the earliest version of the Windows operating system already existed in 1987, few computers used it, and it did not really began to replace DOS until Windows 3.0 in 1990 or 3.1 in 1992.

The Invent simulation would display one or two research teams, each with five to nine members represented by rectangles on the screen in a 3 by 3 arrangement, with varying patterns of social relations represented by lines connecting some adjacent rectangles. If there was one team, it represented researchers at Jet Propulsion Laboratory, and the second team represented researchers at the comparable Ames Research Center. In each run of the simulation, the students would make a few choices, such as the size and social structure of the team, then enter a word like SPACESHIP for the teams to discover. The student could set different characteristics for JPL and Ames, so the simulation run would be a contest to see which team got the full word first, rather like a laboratory experiment in which independent variables could be compared in their effect on the dependent variable.

This simple program gave the student a choice of seven different social structures for each team and four different divisions of labor, letting different subsets of team members discover versus transmit letters of the target word. There was also a rudimentary cultural alternative. Under the default setting, when a team member got its turn, it would prepare either to discover or transmit a letter, and the discover option make it guess one of the 26 letters of the alphabet at random. But letters are not in fact used with equal frequency in the English language, so the program offered a biased alternative, in which letters were guessed with the frequency that each letter tended to appear in English. In the reference source I used, E appears 12.4% of the time, compared with roughly 0.1% for J, Q, or Z (Friedman 1918). Note that the 9-letter word SPACESHIP has only 7 different letters, because both S and P appear twice, thus reducing the effort to discover letters to that of a shorter word like JUPITER having no repeated letters. For most problems, following this biased strategy was more effective, but not for some problems, reflecting the functionality but imperfection of cultures more generally.

As simple as this game-like simulation is to describe, the computer had to work very hard, as each simulated team member required many turns of guessing letters, communicating them to socially connected team members, and assembling the letters into the word. The algorithm for guessing a letter was based on a random number generator, and some key letters might not be guessed for a long time. Where in the social network of the team key letters appeared, could also influence how long it took them to find their way together into a syllable of the word. That meant that repeating the simulation with the same settings would yield somewhat different results.

For an experiment, I modified the program so it would run until all members of a nine-person team had the word ROCKET, then repeat the simulation a number of times, saving all the data on how many turns it took each member to get the word. For purposes of having a clear experiment, the selected social network was a line, representing each team member as a letter of the alphabet:

A – B – C – D – E – F – G – H – I

Table 2.1 shows results for four phases of this experiment. The nine team members were all equally able to discover and communicate, but Scientist A and Scientist I were connected in this linear social network to only one other team member. Thus, on the basis of simple theory, we might predict that they would be the last to get the full word ROCKET, while Scientist E in the middle would be the first. But that is not what we see. The first to get the word was Scientist H, who did so on turn 364. In second place was Scientist B who got the word on turn 412. Scientist E did not get it until much later, on turn 528. Do we now adopt a new theory, that in a social network of this linear shape, the team members just inside the ends have the advantage?

The first column of figures in Table 2.1 reports results from just one run of this simulation, and given how many turns it takes for the simulation to end, there is much room for random variation. The second column of figures reports the averages from 10 runs, and shows a pattern much closer to our original theory. The scientists in the middle of this social network do seem to have an advantage over those at each end, presumably because they can more quickly benefit from discoveries made near both ends. The patterns stabilize as the numbers of runs increase, and the fourth column

Table 2.1 Turns to guess the word ROCKET in nine-member linear teams

Simulated agent	1 simulation run	10 simulation runs	100 simulation runs	1,000 simulation runs
Scientist A	500	614	531	535
Scientist B	412	534	480	482
Scientist C	624	477	438	452
Scientist D	514	470	443	433
Scientist E	528	470	421	430
Scientist F	452	475	415	432
Scientist G	532	496	420	447
Scientist H	364	496	451	477
Scientist I	481	553	525	533

reporting the averages from 1,000 runs, is very nearly symmetrical 535 turns for Scientist A and 533 for Scientist I, just as our original theory would have predicted.

Students would immediately understand that the findings in Table 2.1 were further confirmation of the need for taking large samples from populations, in order to achieve statistical significance. Yet the results also illustrate the nature of chaos and complexity, the likelihood that any one sample will be atypical, and the wandering path likely to go from start to finish in a computer simulation. While extremely simple, this simulation provides a starting point for the development of complexity.

The sociology of science and sociology of technology are well-established fields, and patterns of communication among individuals have been shown to be key factors in producing and promoting innovations. When the simulation compares two laboratories simultaneously, or is run multiple times, it illustrates the fact that a given discovery or invention can be made roughly simultaneously by two or more independent groups or individuals (Merton 1973). This was true for the discovery of the planet Neptune and the invention of the multistage, liquid-fuel rocket, which makes space probes possible. Neptune was predicted by two separate men, and arguably the modern rocket was invented three times independently by Tsiolkovsky, Goddard and Oberth (Grosser 1962; Bainbridge 1976; 1985).

Some sociologists completely reject the idea that individuals do the inventing (Gilfillan 1963). Instead, they suggest, pieces of ideas flow around in the society until the right ones happen to come together. Then an invention is born, and it does not matter which person happens to take credit for it. It was really a result of the whole culture and the set of communication channels that comprise the society. This may be too extreme. But there is no denying that a sociological approach to science and invention, stressing patterns of communication, can help us understand much about these creative processes that are so crucial to our society (Ogburn 1922, 1964; Gruber and Marquis 1969; Crane 1972; Katz and Lazarsfeld 1955; Katz 1960; Rogers 1960). Clearly, many inventions have been the product of teams rather than of individuals alone, and no single person can invent or discover something new without building upon much previous work done by others.

2.2 The Moons of Mars

A series of three theory-based simulations in *Sociology Laboratory* used the exploration and colonization of Mars as a metaphoric context for understanding how humans create and sustain cooperative relations among them, thus creating communities in our real world. Three stages in this educational evolution of society were acted out on Phobos, Deimos and Mars. In Greek mythology, Mars was the god of war, and his sons, Phobos and Deimos, had names expressing horror and terror, or fear and panic. In modern English, Phobos is cognate with *phobia*, and the simulation set on the moon Phobos depicts a situation in which people cannot trust each other. There is room to doubt whether Deimos is cognate with *democracy*, but the simulation set on the moon Deimos explores the evolution of cooperation. The Mars simulation imagines a society becoming large enough that people must learn to trust categories of people, rather than rely entirely upon personal friendships, illustrating a sociological theory of religion, in which a god like Mars is an abstraction from a mental model of reality.

The first of the three simulations in *Sociology Laboratory* that depicted the colonization of Mars was a rather metaphorical replay of the first Apollo lunar mission by three men: Neil Armstrong, Edwin Aldrin and Michael Collins. It imagined that three female astronauts flew the first mission near Mars, actually landing on its moon Phobos: Alice Armstrong, Betty Baldrin, and Connie Collins. They can be represented as A + B + C = team, but the psychological stress of the long voyage, plus a technical glitch upon landing that prevented their spacecraft from returning them to Earth, destroyed their group cohesion, pitting each against the other two in a game of survival until a rescue ship could arrive many months later. So each grabbed resources and rushed away to a different crater to brood about rescue and plot against the other two.

The real point of this simulation was not to tell an adventure story, but to place in a dramatic context a very fundamental lesson about multi-agent systems. Each of the three astronauts was represented by a set of memory registers quantifying the vital resources they possessed and their mental models of the resources that could be obtained from their two opponents. The simulation would go through a series of turns. On each one, the astronauts would assess the situation and decide which action to take. There were two kinds of resources: supplies and energy. Each astronaut had a moon buggy with which she could sneak over to the crater where another astronaut had stored resources and steal either supplies or energy. The buggy would need to be empty to have room for supplies, but would need to carry chargeable batteries with which to steal energy. Thus, each turn would require astronaut A to make two decisions: (1) prepare the buggy for stealing supplies or energy, (2) raid the resources either of astronaut B or astronaut C.

As crude as this set of contingencies may seem, it reflects a standard tradition often called *Learning Theory* or *Behaviorism*. In criminology, Ronald Akers had written at length about Learning Theory, arguing that people will commit crimes to the extent they believe they will benefit from doing so, the rewards greatly outweighing or being far more likely than the punishments, largely on the basis of their past experience (Akers et al. 1979; Akers 1985). George Homans, a leading Behaviorist in sociology, had stated these principles abstractly: “For all actions taken by persons, the more often a particular action of a person is rewarded, the more likely the person is to perform the action. The more valuable to a person the result of his action is, the more likely he is to perform the action (Homans 1974).” This perspective is widespread across the social and cognitive sciences, under a variety of other names, including utilitarianism and pragmatism. Morality features in it only as the framework for estimating some of the costs of performing an action, such as explicit punishment or the more subtle loss of cooperation from any other agents in the social system who are harmed by one’s actions.

Each raid costs the astronaut 10 units of energy to operate the moon buggy. The payoff from the raid may be influenced by the skill of the astronaut, if the student running the simulation wants to apply this optional variable. The payoff is always influenced by random factors, popularly called “luck” and determined by a random number generator in the computer program. Another determinant of the payoff is how much of the desired resource the victim has, because it is quicker to grab a buggy full from a large stash, although it can take many raids to totally exhaust an opponent’s resources. Also, each raid was like a battle between attacker and defender, never wounding an astronaut but including the relative energy possessed by one astronaut compared with the other, among the determinants of how much could be gained on a raid. So despite the simple premise of the Phobos disaster, the contingencies of the simulation provided a somewhat complex background for the astronauts’ decisions. Table 2.2 shows the data from Armstrong’s memory registers of the first 10 turns, both before and after each raid by Armstrong.

RAID 1: At the very beginning, Armstrong has 1000 units of energy and 250 units of supplies. Which will she raid for, energy or supplies? The standard rule is that she will raid for whichever she has less of. Thus she goes for supplies. Before completing any raids, Armstrong has no experience on which to predict what loot she can get from Baldrin or Collins. The simulation placed 10,000 in all the expectation registers at the beginning of a run, a number far higher than could actually be gained, representing not only extreme optimism but also assuring that after the first run each of the three agents would have the beginning of realistic experience on which to base decisions. She happens to choose Baldrin, and she gets 131 units of supplies. Thus, at the end of the raid, she has 381 units of supplies, the 250 she started with plus the 131 stolen in the raid. She places the number 131 in her “supplies expected from Baldrin” memory register; this is what she might expect to get from her on a second such raid. Notice that her stock of energy goes down, from 1000 to 990, because her moon buggy used up 10 units on the raid.

RAID 2: Somebody must have raided Armstrong for energy, because her stock of it has gone down to 961. But she still has less supplies than energy. So, she will again raid for supplies, but from Collins because the “supplies expected from Collins” register still has the unrealistic 10,000 in it, much greater than the 131 expected from raiding Baldrin. So, Armstrong raids Collins and happens to get 143 units of supplies. Armstrong puts the number 143 in her “supplies expected from Collins”

Table 2.2 Ten interactions between one astronaut and two others

	Armstrong's stock of		Energy expected		Supplies expected	
Turn	Energy	Supplies	From B	From C	From B	From C
1 Start	1,000	250	10,000	10,000	10,000	10,000
1 End	990	381	10,000	10,000	131	10,000
2 Start	961	381	10,000	10,000	131	10,000
2 End	951	524	10,000	10,000	131	143
3 Start	951	524	10,000	10,000	131	143
3 End	941	693	10,000	10,000	131	169
4 Start	793	693	10,000	10,000	131	169
4 End	783	780	10,000	10,000	131	87
5 Start	762	678	10,000	10,000	131	87
5 End	752	806	10,000	10,000	128	87
6 Start	730	806	10,000	10,000	128	87
6 End	834	806	114	10,000	128	87
7 Start	714	806	114	10,000	128	87
7 End	754	806	114	50	128	87
8 Start	733	706	114	50	128	87
8 End	723	830	114	50	124	87
9 Start	699	712	114	50	124	87
9 End	816	712	127	50	124	87
10 Start	705	712	127	50	124	87
10 End	835	712	140	50	124	87

memory register. Now she has 524 units of supplies. Again, she spent 10 units of energy to make the raid, dropping from a stock of 951 to 941 units.

RAID 3: Nobody raided Armstrong this turn, so she still has 951 units of energy This is more than the 524 units of supplies she has. So, she will raid for supplies again. Armstrong expects 131 units from Baldrin, and 143 units from Collins, so she raids Collins again. This time she does even better, getting 169 units of supplies. Armstrong puts 169 in her "supplies expected from Collins" memory register. Her stock of supplies increases by 169, from 524 to 693. Her energy goes down by 10, to 783.

RAID 4: Again, Armstrong has been the victim of a raid, losing energy. But she still has less supplies, and Collins still looks like the best victim for her raid. But this time she gets only 87 units from Collins. Perhaps Collins is beginning to run out. Armstrong puts 87 in her "supplies expected from Collins" memory register.

RAID 5: Somebody has stolen energy from Armstrong, but not much, leaving her with 762 units. Somebody else stole over a hundred units of supplies, however, bringing her down to 678 units. Thus, once more she must raid for supplies. She still expects 131 units from Baldrin, the amount she got from her in the first raid. She got only 87 units from Collins the last time she raided for supplies. So she switches back to Baldrin, getting 128 units, which will ensure her next supply raid will also be against Baldrin.

RAID 6: Another small loss of energy and the big gain of supplies she got in her fifth raid gives Armstrong 730 units of energy and 806 units of supplies. For the first time, Armstrong has less energy than supplies. Therefore, she will raid for energy, and gets 114 units from Baldrin, while using up 10, for a net gain of 104, from 730 to 834. Her stock of supplies does not change.

RAID 7: Another loss of energy makes an energy raid imperative for Armstrong. Not having tried Collins for energy previously, she does so. But she gets only 50 units.

RAID 8: A bad raid cost Armstrong 100 units of supplies, and she now has less supplies than energy. Guided by her expectations, based on past experience, she raids Baldrin and comes away with 124 units.

RAIDS 9 and 10: Hit again and again by energy raids, Armstrong keeps raiding for energy herself. Baldrin continues to be a good source of energy, compared to the paltry 50 units last gained from Collins.

While partly determined by chance, the outcomes of the simulation are greatly determined by the settings the student selected at the beginning, such as giving Armstrong 1,000 units of energy but only 250 of supplies. Also, two initial decisions reflect different theories of human memory:

1. STANDARD: For expectations, remember just the result of the last raid.

1. ALTERNATE: Remember an average of the last raid and the prior expectation.

2. STANDARD: Raid the person you expect to get more of the desired reward from.

2. ALTERNATE: Select victim from expectations by proportional probability.

The first option decides whether behavior depends only upon the most recent experience, or also factors in earlier experiences. The first time Armstrong raided Collins for supplies, she got 143 units, so under both the standard and alternate rule, she would expect 143 units from Collins next time. On the second raid she gained 169 units, so under the standard rule, that became he updated expectation. The alternate rule would take an average of 143 and 169, which is 156. On her third raid of Collins she got only 87 units of supplies. So the standard rule updated the expectation to 87. But the alternate rule would take the average of 156 (the previous expectation) and 87 (the new result) which is about 122. The alternate rule emphasizes the most recent result, but moderates it with some memory of the past.

The second option introduces some sophistication of a different kind into the equations describing human behavior. Thinking in terms of ordinary economic payoffs, a higher profit should always be preferable to a lower one, so Armstrong should indeed always raid the other astronaut for which she had the highest expectation. But suppose the drop in the result of raiding Collins for supplies from 169 to 87 was a fluke, and next time she might have gotten much more than from Baldrin? The alternative for this second option actually expresses a standard theory in behavior psychology, Richard Herrnstein's *matching law*, described thus by Wikipedia: "a quantitative relationship that holds between the relative rates of response and the relative rates of reinforcement in concurrent schedules of reinforcement. For example, if two response alternatives A and B are offered to an organism, the ratio of response rates to A and B equals the ratio of reinforcements yielded by each response (Herrnstein 1961)."[1] This kind of thinking influenced Hernstein's Harvard colleague, George Homans, whom we quoted above as theorizing about the likelihood that a person would take a particular action, rather than believing that actions were rigidly determined.

When Armstrong had supply expectations of 131 for Baldrin and 87 for Collins, perhaps she would not have a 100% chance of raiding Baldrin next, but a probability of 131/(131 + 87) or about 60%. The matching law remains controversial, especially in its specific calculation, but the general idea is certainly valid: Humans benefit from occasionally departing slightly from what seems the optimal course of action, thereby gaining improved information about the actual contingencies they face in this complex world.

Students were encouraged to try the Phobos simulation several times, both to experience the chaotic results of social relations influenced by random factors, and also to experiment with alternative decision rules or starting conditions. However, moving on to the Deimos simulation placed the exercises in a much richer conceptual context. It imagined that a dozen astronauts were sent to Deimos to build a base from which Mars itself would be colonized. Demoralized by what had happened to the Phobos expedition, upon landing on Deimos, they frantically grabbed resources at random and set up personal hideaways in separate craters. To survive for long, they needed to trade resources, but were reluctant to come near each other, in fear of violence, so they would make an agreement on radio, then place what they were willing to give in one location, then go to another to take what they were given in return. How could they learn to trust each other?

At the time *Sociology Laboratory* was published in 1987, the use of computer simulations to assess theories of social and economic interaction was both relatively new and extremely popular. In 1984, political scientist Robert Axelrod had published *The Evolution of Cooperation*, using computer simulation to explore the implications of a conceptual experiment called the *prisoner's dilemma* (Axelrod 1984). Similar to a strategy choice presented by a computer game, it was described thus in 1969 by psychologist Kenneth Gergen:

> Two men are suspected of having committed a crime together. They are placed in separate cells, and each is informed separately that he may choose to confess or not, subject to the following stipulations: (1) If both independently choose not to confess, they will receive only moderate punishment. (2) On the other hand, if one chooses not to confess and the other simultaneously confesses, the confessor will receive the minimal sentence while his partner will be given the maximal sentence. (3) If both choose to confess, they will both be given heavy sentences (Gergen 1969).

This is an interesting dilemma, because it is in the interest of each prisoner to confess, but if both confess they wind off worse than if they did not confess. For years, social psychologists had conducted many experiments a little like our simulation with real people interacting in the laboratory to see if cooperation could emerge

[1] en.wikipedia.org/wiki/Matching_law.

under conditions lacking most of the features we normally find in society. Often, cooperation has developed, even out of very simple social interactions (Sidowski et al. 1956; Kelley et al. 1962). Robert Axelrod's book reports the results of a computer simulation in which the prisoner's dilemma was played over and over again. His computer program included many different strategies for playing the game, strategies that actually came from a public contest he ran. Under many circumstances, the winning strategy had been submitted to the contest by Anatol Rapoport, and is the fourth in the following list that were programmed into the Deimos simulation:

1. Always give, regardless of what the other person does.
2. Never give, regardless of what the other person does.
3. Behave entirely at random, giving about half the time.
4. Give at first, then do what the other person did last time.
5. Same as above, except fail to give 20% of the time.
6. Learn whether giving or not giving is more rewarding.

The way the Deimos simulation ran, the dozen simulated people did not have very complex artificial intelligence, but would do 20 trades with another simulated person, following one of these rules, of which numbers 4, 5 and 6 involved simple machine learning. Students were able to compare strategies, and see what happened when simulated people were following different strategies.

The fact that Rapoport's strategy won in Axelrod's contest does not prove it is better, because much depends upon the complexity of a simulation, the size of the population of artificial agents, and the extent to which sub-groups can form within the simulated population. This kind of conceptualization has been applied in sociobiology, the study of the evolution of animal behavior, in which species evolve somewhat independently yet in interaction with each other. A prominent example is the work of John Maynard Smith - whose unconventional last name is the two words Maynard Smith - in his 1982 book *Evolution and the Theory of Games* (Smith 1982). Yes, most animals in a particular environment may be vegetarian herbivores, but some are carnivores who eat the herbivores. Thus in human societies, cooperation may evolve within each distinct social group, but it may be to the advantage of members of one group to exploit the members of another - if they can get away with it!

2.3 Social Complexity

Since a highly publicized conference held at Dartmouth College in 1956, computer scientists and their advocates have been promising great progress through "artificial intelligence," a set of concepts and programming schemes that could allow computers to equal or even surpass the human ability to understand their surroundings and decide upon the best actions to achieve even lofty goals (McCorduck 2004). Yet "AI" as it is often called could also be defined as a set of metaphors that have too often been used to misrepresent computational methods and accomplishments. This has not merely meant that the values of some methods have been exaggerated, but also that other goals the methods might accomplish have been underestimated. These issues are so vast we can consider only two relevant aspects of them here: (1) categorization of computational methods, and (2) adequacy of solutions for problems.

Two of the programming methods that have experienced somewhat erratic histories within AI are *neural networks* and *genetic algorithms*. Both can be conceptualized as subcategories of *machine learning*, and clearly the Phobos and Deimos programs were examples of machine learning that seemed similar to neural networks. Neural networks typically imagine a network of brain neurons connected to each other, in which each connection is represented in computer memory, often with a number described as a *weight* (Bainbridge 1995a). Multiple lines of the network, each with a weight that can be changed, connect stable points called nodes. Data abstractly representing a question enters the network's input, and the pattern of weights between the nodes determines the answer produced as the output. The rest of the system evaluates the answer, and feeds back information about how well it responded to the question, adjusting the weights in a way likely to improve the behavior of the network over a series of iterations.

In their 1969 book, *Perceptrons*, AI pioneers Marvin Minsky and Seymour Papert raised criticisms of neural networks, that seemed to say they could never solve some even rather simple problems (Minsky and Papert 1969). As its Wikipedia page reports, "This book is the center of a long-standing controversy in the study of artificial intelligence. It is claimed that pessimistic predictions made by the authors were responsible for an erroneous change in the direction of research in AI, concentrating efforts on so-called 'symbolic' systems, and contributing to the so-called AI winter. This decision, supposedly, proved to be unfortunate in the 1980s, when new discoveries showed that the prognostics in the book were wrong."[2] My reading of *Perceptrons*, and interactions both literary and in person with Minsky, did give me the impression that it served to strengthen the very different "symbolic system" approach, often called *rule-based reasoning*, in which AI operates through the equivalent of deductive systems, like those based on axioms in mathematics or symbolic logic. Later in life, Minsky clearly believed that many forms of AI had value, but each had distinctive advantages under certain circumstances (Minsky 2008).[3]

Genetic algorithms also consist of structures of computer memory registers, but conceptualized in terms of a population of independent biological organisms rather than a network of biological neurons inside a single brain (Holland 1975). One of the programs in *Sociology Laboratory* illustrated genetic algorithms. Called Evolve, it simulated sociobiology of glow-worms on Neptune's moon, Triton. Whatever decisions the student enters, such as how quickly the worms die of old age, the program begins with a population of males and females that differ only in the gene determining gender. The memory register representing the gender gene contains 1 for a male, and 2 for a female.

During a run, the worms may mate, and each child will have a random combination of parents' genes, but with some chance of mutation for each gene. The glow worms

[2]en.wikipedia.org/wiki/Perceptrons_(book), accessed December 2017.

[3]www.terasemjournals.org/PCJournal/PC0303/mm1.html, accessed December 2017.

are somewhat like terrestrial fireflies. A male travels around until he comes near a female, toward whom he flashes his glow light once. If she responds with one flash, they mate, producing one child. There is a flash gene, represented by a memory register for each worm containing a 1 for one flash, or a 2 for two flashes. At the beginning, all these memory registers are set at 1, so all the couples flash once at each other, and mate. But a child may have a mutated 2 in this memory register, which means 2 flashes are emitted and required for mating. Some time into a particular run, there may be enough worms with this gene so they can reproduce. This represents the evolution of a new species, reproducing within a separate gene pool.

Two other genes determined aggressive behavior by the worms, in an intentionally role-specific interaction with gender so that students could see how complex social patterns might emerge from very simple simulations. On Triton, the gender of a worm determines a good deal of behavior, males moving around while females remain in place, and males capable of attacking other males but not attacking females. Each begins with a pacifist 0 in its memory register, but mutation might motivate aggression in different ways for the two genders:

> Attack gene: A male with 0 has no tendency to attack, 1 represents the ordinary attacker who can eat a male with 0 for this gene, and 2, the superattacker who can also eat those with 1. Females will carry these genes, too, but only their male offspring will exhibit the attacking behavior.
>
> Trap gene: A female with 0 will ignore males of the other species; a female with 1 will eat them if they try to mate. Males carry the trap gene, but only the female offspring exhibit the trapping behavior.

The simulation imagined a large rock, inhabited by 50 males and 50 females. Offspring swim away from the rock, if they cannot find space on it freed up by the death of one of the original 100 worms. Students could run the simulation in a series of steps, changing the mutation probabilities for the genes separately, trying to create a population rife with violence, or two peaceful species, or some more dynamic outcome.

Setting aside this drama about alien lifeforms, the $100 \times 4 = 400$ memory registers could easily be conceptualized in other ways, for example as neurons in the mind of a single creature. Thus, I tend to think of neural nets and genetic algorithms as metaphors for conceptualizing complex memory structures, rather than really being very different kinds of artificial intelligence.

2.4 The Gods of Mars

The Mars simulation imagined a colony of 24 people who performed different economic functions and would gradually learn which other colonists to interact with to obtain specific resources. The complex algorithm governing their artificial intelligence could be described as a development based on conventional social statistics, or as an innovative form of neural network artificial intelligence, which was later used for the 1995 journal article, "Neural Network Models of Religious Belief (Bainbridge 1995b)." While citing this article in 2005, M. Afzal Upal commented, "Although computer simulation has become a relatively accepted technique for studying social theories, it has rarely been used to study religion (Upal 2005)." Subsequently, a much expanded multi-agent version of the Mars simulation, with thousands of colonists, was programmed for the 2006 book, *God from the Machine: Artificial Intelligence Models of Religious Cognition,* and will be used again in the following chapter (Bainbridge 2006). In 2008, independently of my work but citing Upal, James Dow published results from a simulation suggesting that "a central unifying feature of religion, a belief in an unverifiable world, could have evolved along side of verifiable knowledge," rendering religion an evolutionary adaptation (Dow 2008).

We can imagine that a colony arrives on Mars with a large supply of technological tools but without the raw materials to survive more than a few weeks without resupply. Solar power devices could produce the needed energy, operated by some of the colonists, but water, food, and oxygen would need to be produced from local Martian resources. At the beginning of a run of the 1987 simulation, the student would decide how many different occupational categories the 24 colonists would be divided into - whether 2, 3 or 4 - and which category would perform which of four jobs. For our example here, the colonists were divided into four categories, each with one job:

1. Operate the solar power units to produce energy.
2. Use energy to melt ice to produce water.
3. Use water on Martian soil to produce food.
4. Combine energy and water to produce oxygen.

As in the Phobos and Deimos simulations, the colonists would go through a number of turns. Periodically, they would do their jobs, and spend the rest of their time trading with each other. A colonist would seek whichever resource the colonist has less of, and prepare to give in return whichever resource the colonist had most of. As in the Phobos simulation, the colonist would select another colonist to exchange with, but by peaceful trading rather than theft. If the other colonist would gain through the proposed trade, then it takes place, and the colonists learn that a trade like this is worth considering in the future.

The algorithm is complex, and operates in two stages. First, a colonist decides whether to assume there are 2, 3 or 4 categories of other colonists. Second, within that hypothesis, the colonist must guess which category will provide the desired resource. In order to make this computable in an educational program on a primitive computer, the colonists were identified by the numbers 1 through 24, and assigned jobs in a cycle. In this example, producing energy was performed by colonists 1, 5, 9, 13, 17, and 21. Producing oxygen was performed by colonists 4, 8, 12, 16, 20, 24. The artificially intelligent minds of the colonists used this system to classify each other, but had sets of memory registers for the alternative hypotheses that there were 2, 3, or 4 categories. Over time, they would learn relatively efficient but not entirely perfect understanding of the actual social structure. The program also permitted an experimental complexity of great significance: The 24 artificial intelligence agents could share information about their past experience interacting with different groups,

Table 2.3 How the Mars simulations displayed one agent's energy memory

Theories about the categories	Question 1: which theory	Question 2: which group within a theory			
		Group #1	Group #2	Group #3	Group #4
2 Groups	512	499	4	5	4
3 Groups	148	30	30	87	1
4 Groups	340	267	2	68	3

Table 2.4 The actual content of one agent's energy memory

Theories about the categories	Variance across 4 Groups	Consequences of 40 turns of the simulation			
		Group #1	Group #2	Group #3	Group #4
2 Groups	91875.50	706	5	7	6
3 Groups	26588.19	155	157	455	6
4 Groups	60995.25	608	5	155	6

thereby learning more quickly through cooperation, and developing the equivalent of a Martian culture.

At any point, students could select one of the 24 agents and inspect a slightly simplified version of its memory concerning any one of the resources. Table 2.3 shows what one agent had learned about obtaining energy, after 40 turns of one run of the simulation. On any turn, the agent first selects a theory, using a random number generator that will give 512 chances out of 1,000 of selecting the theory there are 2 groups, 148 out of 1,000 of selecting the 3-group theory, and 340 out of 1,000 of selecting the 4-group theory. If the agent selects the 2-group theory, then it will have a probability of 499 out of 512 in selecting group 1 among the 2 groups, and a minuscule 4 out of 512 chance of selecting group 2 of 2 groups. If the agent selects the 4-group theory, there will be a chance of selecting group 1 of 267 out of 340.

The 15 cells containing numbers in Table 2.3 do represent 15 memory registers, but the display was programmed to simplify somewhat, for sake of comprehension. The actual contents of the memory registers are shown in Table 2.4. The simulation began by putting small randomly-chosen numbers in the 12 memory registers for question 2, then calculating the statistical variance across each set of 4 numbers. On those turns in which the agent sought energy, the amount of energy gained would be added to the cell for each of the three theories to which the exchange partner belonged, as categorized by that theory. Then the variances would be recalculated.

Notice that the agent favors the theory that there are two groups, putting the theory that there are four groups in second place. But there actually were four groups, not two. This is an important point. If the goal in creating artificial intelligence is to find the objectively optimal solution to a problem, then this simulation is NOT the best kind of AI. However, humans often select sub-optimal solutions to problems, so simulations intended to model human behavior may often be intentionally sub-optimal to serve that theoretical goal.

In a different study, using a standard form of neural network, I had explored how humans might develop prejudge against all members of an unfamiliar ethnic group, if some members of that group caused problems in social interaction (Bainbridge 1995c). Many of the agents would learn to reject members of that group, at least at first, and only with difficulty learn to be more "discriminating," rejecting only members of the group that possessed other characteristics warning that interaction with them might be costly. This was explicitly a simulation based on Gordon Allport's cognitive efficiency model of prejudice (Allport 1954).

In the Mars simulation, if the conditions were set so that agents could communicate with each other, and such that some members of the society came to the view that there were four groups, this more accurate theory could progressively become more popular until it was the cultural norm. But notice that the memory was constructed so that the agents could believe in the existence of three logically contradictory possibilities: (1) the third of two groups, (2) the fourth of two groups, and (3) the fourth of three groups. This was built into the program specifically to represent a theory of religion that Rodney Stark and I had been exploring in publications dating from 1979 to 1987, culminating in our axiom-based book, *A Theory of Religion* (Stark and Bainbridge 1997).

It is not necessary to go into that theory deeply here, as I did so in *God from the Machine* in 2006. But the metaphor is easy enough to understand. If there are really three categories of human beings, a deity is like a human but does not belong to a category that contains any human. In number theory, we can contrast real numbers from imaginary numbers: the square root of −1 is +1, but the square root of −1 is not −1, because that is an alternate square root of +1. The square root of −1 is i, the archetypical imaginary number.

Divide each of a series of integers by 3, and record the remainder: 1, 2, 0, 1, 2, 0,... Those can be called group 1, group 2, and group 3. Group 4 consist of all integers for which the remainder of dividing by 3 is 3. Do any such integers exists. Well, yes and no. No, there are no natural numbers that have that property, those numbers used in ordinary counting and measurement. But we can imagine a supernatural number, call it s, for which it is true that the remainder of dividing s by n equals n. The algorithmic procedures represented by the two previous tables can indeed manage supernatural numbers. How, and what do they represent?

The Martians seek not four resources but five: energy, water, food, oxygen, and life. They begin with some amount of each of these resources, and each turn they use up some water, food, oxygen, and life. Through doing work and exchanging with each other, they can replenish water, food, and oxygen, but not life. When life reaches zero, they die of old age. Before that happens, they begin seeking life from each other, just as they may seek any of the other resources. Each agent has a set of 15 memory registers for each of the five resources, learning separately where to seek each of them. Table 2.5 shows what one agent's life registers might show after

Table 2.5 The actual content of one agent's life memory

Theories about the categories	Variance across 4 Groups	Consequences of many turns of the simulation			
		Group #1	Group #2	Group #3	Group #4
2 Groups	10.24	0	0	7	6
3 Groups	5.76	0	0	0	6
4 Groups	0.00	0	0	0	0

many, many turns. The agent has interacted many times with other agents, always receiving 0 life. But the agent was unable to interact with the postulated supernatural beings, so the initial, small, random numbers placed in the memory registers were never replaced - not by zeros and not by real numbers.

When faced with the problem of seeking an unavailable reward like eternal life, this agent will never select the theory that with respect to this kind of reward the world is divided into four groups of beings, because the variance across 4 memory registers containing zeroes is zero. The agent will have a 36% chance of deciding to seek help from the fourth group in a world consisting of three groups −5.76/(10.24 + 5.76) - perhaps, if we may imagine, the Christian saints. That leaves a 64% chance of deciding there are in fact just two groups of natural beings in the world, and two groups of supernatural beings, perhaps Satan's demons versus Christian saints, with nearly equal chances of selecting either.

2.5 Deviance and Social Control

Figure 2.1 shows three versions of the on-screen display of another one of the 1987 simulations, called Morals, which was originally set in Chicago but could easily be moved to Mars. It models a community of 190 people, each represented by a small rectangle, connected by a few social bonds, represented by "=" signs. The simulations were programmed to run on the primitive machines of the day, many of which lacked graphics cards, so the image was built up from individual letters of the alphabet or spaces given different colors. We can imagine that over the past three decades the 190 Chicago residents has become fed up with the politics of the city, or have retired with good pensions, and decided to migrate to Mars. The only questionable assumption here is that space technology developed rapidly enough after 1987 to make that possible.

The reason the original simulation was set in Chicago was because it was based on theories of the so-called Chicago School of sociology, in particular Edwin Sutherland's *Differential Association Theory* and Fritz Heider's *Balance Theory* (Sutherland 1947; Sutherland and Cressey 1970; Heider 1958; Curry and Emerson 1970). Differential association theory is often interpreted to mean that our attitudes are shaped by

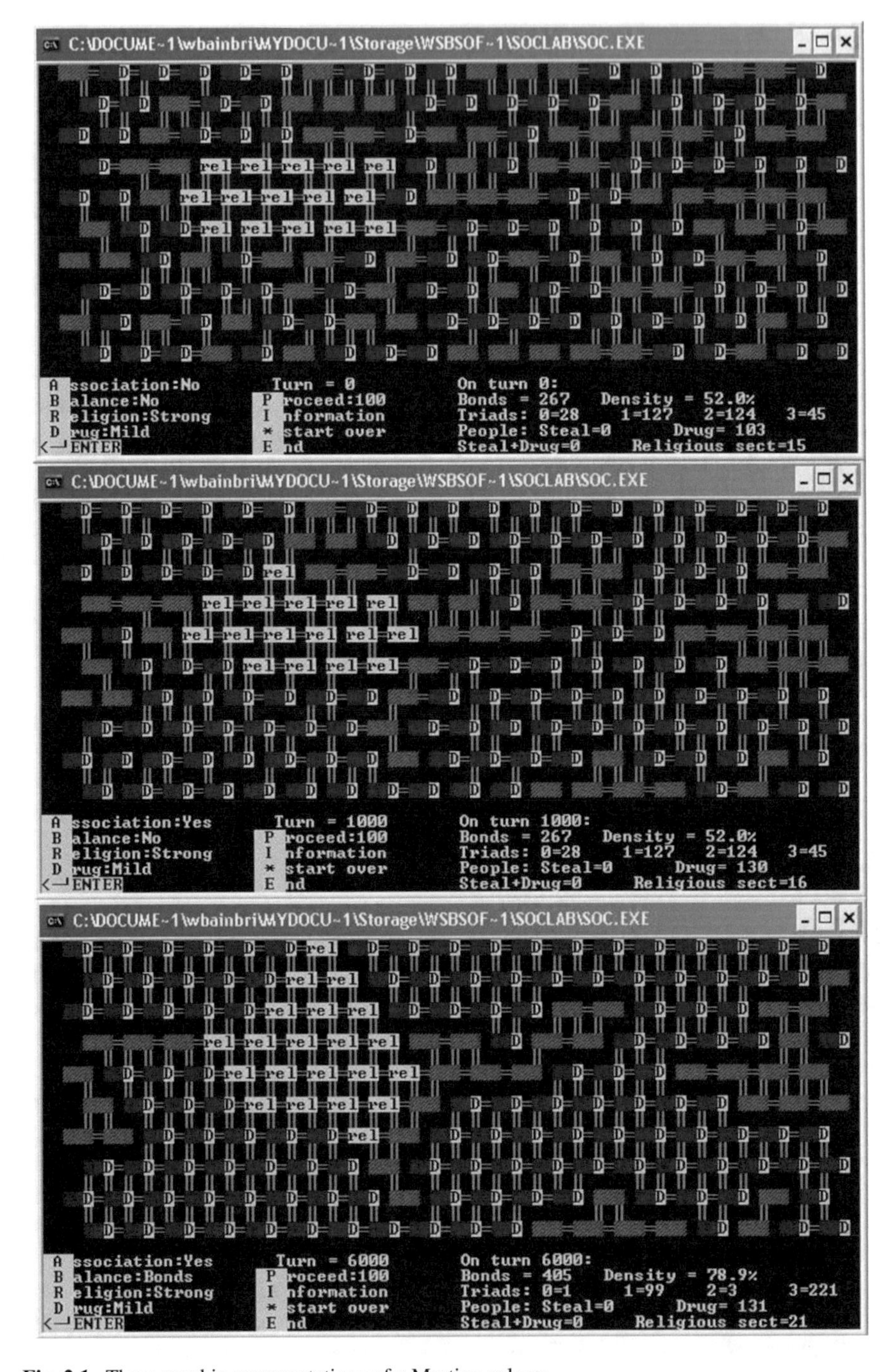

Fig. 2.1 Three graphic representations of a Martian colony

the people with whom we associate. That may be true, but is an oversimplification of what Sutherland proposed. He focused on the communications people receive from each other, arguing that people become deviant if the preponderance of messages they receive endorse deviant behavior. He initially used the word *association* to refer to the mental associations humans have in their minds, that connect ideas and memories. Thus differential association theory can be included within cognitive science, and readily represented computationally.

Similarly, Heider's balance theory can be understood in at least two ways, (1) concerning social relationships, or (2) concerning attitudes. Consider astronauts Armstrong, Baldrin and Collins before they left Earth on their ill-fated journey to Phobos. During their trip, did they develop a friendship triad? Balance theory would set aside everything else to look at their initial relations. Suppose Alice Armstrong and Betty Baldrin had already become friends before launch, and Betty Baldrin was also friends with Connie Collins, but Armstrong and Collins did not know each other well. So, there are these friendship ties: A = B = C but no A = C tie. Suppose the accident on Phobos was only the last of their problems, and they had furiously argued about how to handle the spaceship during their long flight, Armstrong expressing one view, Collins another, and Baldrin undecided. Under normal circumstances, balance theory could consider that the lack of a friendship between Armstrong and Collins was unbalanced, which would have predicted that Armstrong and Collins would have formed a friendship, connecting the trio in a triad. But with disagreements, Baldrin may be forced to break the friendship tie with either Armstrong or Collins.

Cognitive equilibrium requires that a person's beliefs and social attachments needed to harmonize. Having a friend who had different attitudes would cause *cognitive dissonance*, which might impel one individual to convert to the other's viewpoint (Festinger 1957). But balance could also be achieved by breaking off the friendship. Similarly, having a neighbor who shares one's own attitudes, but is not a friend, causes cognitive dissonance. One might resolve that different form of imbalance by building a friendship with the like-minded neighbor. In many applications, the focus is on just a dyad of two people, A and B, and a belief, attitude or action takes the place of C.

The top image in Fig. 2.1 shows the 190 people in a situation in which a concentrated group of 15 are members of a religious sect, represented as [rel]. Some non-members are marked with the letter D like this: [D]. That indicates they share the use of recreational drugs, while the reminder of the Martian colonists are neither religious nor druggy. The text at the bottom right of the top image reports there are 103 drug users. Another simulation option, to add thieves identified [S] for stealing, is not used in this particular run of the simulation, so their total is 0, as is the total for their combination [S D]. The density of social bonds is 52.0%. Prior to starting the simulation, the student would have made decisions such as these.

The religion variable has been set at Strong, which means that people marked [rel] will intentionally develop social relations with their neighbors, in order to convert them. The student then sets differential association theory to Yes, and runs the simulation forward 1,000 turns. In each turn one of the 190 people will be selected at random and will follow differential association theory, essentially taking a vote of all their friends to see what their behavior should be: religious, druggy, or neither. Over the 1,000 turns, the religious group grew from 15 to 16 members, and the druggies gained from 103 to 130. Balance theory was not yet turned on, so there is no change to the density of the social network. Balance theory can be set to affect either social bonds directly, or attitudes, and here was set to bonds. Then the simulation went forward another 5,000 turns.

The religious sect grew to 21 members, the druggies gained only one recruit to 131, and the density of the social network jumped from 52.0 to 78.9%. Another way to measure the change in social structure is in terms of how many of the trios of close neighbors are triads of friends, and this number rose from 45 to 221. Combined here with the three simulations that explicitly model Mars colonization, the Morals program illustrates how standard social-scientific theories can be modelled in computer simulations designed for educational purposes, a topic we shall return to in Chap. 4.

2.6 The Wider Literature

As the previous chapter noted, In his recent textbook, Claudio Cioffi-Revilla has identified computer simulation as one third of the triad he calls *computational social science* (Cioffi-Revilla 2014). Yet studies primarily using simulation methods are rather rare among the articles published in the most prestigious social science journals. One optimistic way to conceptualize this situation is to draw an analogy with the emergence of cognitive science as a distinct discipline decades ago. Largely influenced by work in artificial intelligence by computer scientists, *cog-sci* is rather distinct from cognitive psychology, giving more emphasis to computational neural science, and connecting to computer vision and computer-based natural language processing. It is important to note that cog-sci does not include much if any work from two of the most cognitive of the traditional human sciences, namely sociology and political science, which have always emphasized beliefs and attitudes, which are cognitive functions.

Therefore, if we seek simulation studies that offer ideas and methods relevant to questions related to spaceflight social behavior, the *American Sociological Review* and *Journal of the Royal Anthropological Institute* are not the places to look. Instead, we should start with a specialized online journal, *The Journal of Artificial Societies and Social Simulation* (JASSS). The first issue was published January 31, 1998, and by June 30, 2017, 78 issues had been published, containing book reviews as well as articles.[4] Claudio Cioffi-Revilla was author or co-author of 6 of the publications in JASSS, including a solo article, "A Methodology for Complex Social Simulations." It notes how difficult it can be to model complex social systems and advocates starting with simple models and then "conceptualizing and developing a succession of models with increasing complexity as they approximate the target system (Cioffi-Revilla 2010)." He observes that three different kinds of complexity combine to magnify

[4]jasss.soc.surrey.ac.uk/index_by_issue.html.

the difficulty of the task: (1) inherent complexity of the structure and interaction process of the social system, (2) added complexity associated with multiple levels of analysis, such as from individual to group and over different spans of time, and (3) the complexity that scientists bring to the research themselves, especially if they are a multi-disciplinary team. He illustrates his point with the results of a complex project that has been supported through the National Science Foundation, described thus in its online grant abstract:

> Inner Asia played an essential catalytic role in the history of human and social dynamics in Eurasia. This system produced the largest territorial polity to emerge in the evolution of civilizations (the Mongol Empire of the 13th and 14th centuries) as well as the largest-scale economy (the Silk Road network) before present-day globalization. This project will address fundamental questions in theory, methodology, and data concerning human and social dynamic responses to social and environmental challenges in Inner Asia and the Eurasian world system over the past 4,500 years. The interdisciplinary team includes expertise from the social sciences and computational sciences, including political science and international relations, archaeology, cultural anthropology, physical anthropology, economics, applied mathematics, and computer science. This project also includes international collaborations with experts on Inner Asian archaeology and ethnography as well as European experts on computational models of empire dynamics and Silk Road network simulations. The researchers will use diachronic data from written sources and from three archaeological projects situated along a north-south transect spanning the Mongolian steppe zone as well as object-oriented, agent-based computer models that build upon and extend extant computational social science models to generate the emergence of multi-scale networks over space and time.[5]

This project is highly relevant to this chapter in two ways. First, to this point the chapter has offered a series of rather simple simulations, programmed in the classic style of three decades ago, and Cioffi-Revilla inspires us to become more ambitious. One way this book seeks to approach that goal is through exploration of very complex virtual worlds in the concluding chapters. Second, simulation of Inner Asia involves creating a realistic model of a complex historical social environment, yet we could well imagine that mature Martian colonies might interact in ways rather similar some centuries in the future. Indeed, ancient Eurasia had complex but difficult transportation systems such as our solar system might experience once all habitable places had their own semi-autonomous cultures.

Twice JASSS has published studies of itself, in 2009 and 2017, based on quantitative analysis of its publications (Meyer et al. 2009; Hauke et al. 2017). Co-citation analysis reported in the more recent study placed articles in eight initial categories: (1) reciprocity, (2) evolution of cooperation, (3) simulation of science, (4) marriage models, (5) opinion dynamics, (6) standards, (7) methodology, and (8) tools and platforms. The first two were strongly connected to each other, and the last two also had some connection. The "marriage models" category actually did simulate the process by which humans form families, rather than being a metaphor for multidisciplinary marriage of scientific fields. The studies reported earlier in this chapter would seem primarily to fit these three: reciprocity, evolution of cooperation, and opinion dynamics. In its self-study, JASSS explored several modes of analysis beyond mere

[5] www.nsf.gov/awardsearch/showAward?AWD_ID=0527471, accessed December 2017.

categorization, yet it lists the articles in each category, which is a convenient starting point for seeking studies that could be re-interpreted in ways relevant to extraterrestrial societies. Indeed, the citation analysis goes far beyond JASSS, listing the most frequently cited publications in each category, even if they were published elsewhere, thereby providing a convenient introduction to computer simulation of social systems more generally, although many of the external citations are to publications that provide theory rather than performing simulation.

The Morals simulation described in the previous section was clearly a very primitive example of an opinion dynamics program, as well as modeling dynamic social relations that could be affected by opinion differences. JASSS counted six of its most highly cited early articles in this category. Here are their titles, which hint at the scope of this category (Davidsson 2002; Hegselmann and Krause 2002; Deffuant et al. 2002; Urbig 2003; Deffuant 2006; Urbig et al. 2008):

"Agent-Based Social Simulation: A Computer Science View"

"Opinion Dynamics and Bounded Confidence: Models, Analysis and Simulation"

"How Can Extremism Prevail? A Study Based on the Relative Agreement Model"

"Attitude Dynamics with Limited Verbalisation Capabilities"

"Comparing Extremism Propagation Patterns in Continuous Opinion Models"

"Opinion Dynamics: The Effect of the Number of Peers Met at Once"

Many other JASSS articles are worthy of attention. A classic problem in computer simulation of social interaction, especially multi-agent systems that model ideological diversity, is what might prevent the population from completely adopting an orthodoxy (Kurahashi-Nakamura et al. 2016). At the same time, simulation research has often focused on polarization, the separation of a population into two competing ideological subcultures (Siedlecki et al. 2016). A simulation earlier in this chapter modelled the emergence of three competing subcultures on the planet Mars, and yet the general idea might describe the development of three separate cultures on Earth, Moon and Mars. Each could easily broadcast locally the mass media from each of the others, yet social bonds between them will be weak. In terms of time, travel between Earth and Moon is much briefer than travel between them and Mars, and yet much of the expense is the cost of launching from Earth which has by far the most powerful gravity of the three. Also, while Mars has an atmosphere, it is quite thin, so some of the technological features of everyday life on Mars and the Moon may be more similar to each other than either is to Earth. Thus it is plausible that distinctive factors might shape the culture on each, producing an approximate equilateral triangle of cultural differences.

Colonization of another planet in our own solar system could have very different motivations than expansion of a civilization across interstellar distances. Most obviously, it is conceivable that trade of physical objects could be economically feasible, if limited perhaps to high-value items. For example, admittedly using simplistic examples, Mars could ship gold to Earth in return for computer chips. Without assuming we yet know what interplanetary trade would involve, we can consider analogies with empirical research and computer simulation of international trade on Earth.

Ermanno Catullo noted that a considerable body of literature indicates that, in some industries at least, corporations that engage in international trade are more successful than ones that do not:

> There is strong empirical evidence that export firms have a better performance than firms that do not export (exporter premia): exporters are larger, they are relatively more capital and skill intensive, they present higher total factor productivity and higher value added per worker. The exporter premia may be the result of a self-selection effect: only the most competitive firms are able to export because only these firms can overcome fixed costs to enter foreign market (building a commercial network abroad, adapting goods and services to foreign standard and tastes etc.) and variable costs of exporting as transport costs and travel assurance expenditures. On the other hand, exporting may improve firm performance for multiple reasons: firms may access foreign knowledge and technologies (technological spillovers), exporting may increase sales opportunities and, therefore, disposable resources. Moreover, export firms may become more innovative in order to sustain international competition (Catullo 2013).

His computer simulation models corporate decision-making, somewhat simply built based on careful theory and mathematical analysis, exploring how the cause and effect factors are mutually reinforcing. A company that has a history of making good decisions is in a better position than others to afford the initial fixed cost of exporting, and then is in a good position to benefit from it.

A Chinese team of researchers simulated the related problem of mergers and acquisitions (M&As) in an international context: "The use of overseas M&As is a feasible way of seeking external resources to upgrade a company's global competitiveness and technology innovation ability. However, due to cultural differences, different management patterns and technology gaps, acquiring companies may face huge difficulties in transforming core technology, such as patents. Without a powerful post-merger strategy, acquiring companies are more likely to miss the opportunities to make technology innovation, leading to merger failures (Chen et al. 2016)." A Brazilian team used simulation to explore the economic and social implications of how government taxation was carried out across a wide geographic territory, more or less centralized (Furtado and Eberhardt 2016).

2.7 Conclusion

Claudio Cioffi-Revilla may be correct that computationally sophisticated social sciences can gradually build increasing complexity into their simulations, while retaining the rigor that well-designed simple simulations may possess. However, the later chapters of this book explore another way to add value and achieve continued progress: adding human beings into the equation, and in space-related massively multiplayer online games. Most obviously in the educational and strategy games of Chaps. 4 and 5, elements of programs like the rather antique examples emphasized in this chapter still appear and have intellectual relevance, but within much larger and more diverse virtual environments populated by real people. But similar mechanisms and parameters will appear in the four concluding chapters as well. The following chapter builds on one of Cioffi-Revilla's other points, that complexity may come from multiple levels of analysis, such as from individual to group and over different spans of time. Chapter 3 goes beyond planting tiny colonies on Mars, to interstellar colonization and communication, a far larger scale of complexity than we currently experience here on Earth.

The simple simulations mentioned in this chapter are not only good educational tools, but they also can be effective as tests of the logical coherence of a formal theory. Applied to social behavior in the early stages of a Martian colony, they may also be appropriate. Consider the fact that a very complex manufacturing economy on Mars can be assembled only many years after the colony's establishment. For a long time, all the microelectronics will need to be imported from Earth, because they are feasible to build only with a vast system of component creation and assembly, using raw materials mostly available on Mars but refined in very complex ways. Consider how advanced a Martian economy will need to be before it can produce its own wool, cotton, wood, and even plastic. Many specialized intellectual occupations, from medical diagnostics to college teaching, may for many years remain on Earth and be sent to Mars as information services, rather than requiring the extreme division of labor on Mars that is found in the large population on Earth. Thus, early colonies may be suitable for relatively simple simulations precisely because the real social experiments on Mars must be designed for simplicity if they are to function efficiently and reliably.

A rather different insight relevant for social simulation is the standard literary trope of science fiction, that inhabitants of distant planets may be *alien*, in the sense of strange and exotic, possessing very different cultures from those of Earth. This may be true even if the "extraterrestrials" are remote descendants of Earthling colonists, whose culture diverged due to isolation and different natural conditions. Thus, whether or not we physically encounter other technologically advanced civilizations "out there," it would be reasonable for simulations to explore how a set of very different high-tech civilizations might interact with each other. The simplest model would be a set of separate simulations, one for each civilization, that exchanged information through rather narrow connections, most obviously the situation if contact is entirely remote via radio. Many of the space-related massively multiplayer online games simulate much more intimate interaction, as members of different alien civilizations inhabit the same territory, competing for the same natural resources, often locked in bitter conflict.

References

Akers, Ronald L. 1985. *Deviant Behavior: A Social Learning Approach*. Belmont, California: Wadsworth.

Akers, Ronald L., Marvin D. Krohn, Lonn Lanza-Kaduce, and Marcia Radosevich. 1979. Social Learning and Deviant Behavior. *American Sociological Review* 44: 636–655.

Allport, Gordon. 1954. *The Nature of Prejudice*. Boston: Beacon.

Axelrod, Robert. 1984. *The Evolution of Cooperation*. New York: Basic Books.

Bainbridge, William Sims. 1976. *The Spaceflight Revolution*. New York: Wiley.
Bainbridge, William Sims. 1987. *Sociology Laboratory*. Belmont, California: Wadsworth.
Bainbridge, William Sims. 1991. *Goals in Space: American Values and the Future of Technology*. Albany, New York: State University of New York Press.
Bainbridge, William Sims. 1995a. Minimum Intelligent Neural Device: A Tool for Social Simulation. *Mathematical Sociology* 20: 179–192.
Bainbridge, William Sims. 1995b. Neural Network Models of Religious Belief. *Sociological Perspectives* 38: 483–495.
Bainbridge, William Sims. 1995c. Minimum Intelligent Neural Device: A Tool for Social Simulation. *Mathematical Sociology* 20: 179–192.
Bainbridge, William Sims. 2006. *God from the Machine: Artificial Intelligence Models of Religious Cognition*. Walnut Grove, California: AltaMira.
Bainbridge, William Sims. 1985. Beyond Bureaucratic Policy: The Spaceflight Movement. In People in Space, ed. James Everett Katz, 153–163. New Jersey: New Brunswick (Transaction).
Catullo, Ermanno. 2013. An Agent Based Model of Monopolistic Competition in International Trade with Emerging Firm Heterogeneity. Journal of Artificial Societies and Social Simulation 16(2) 7: 1.
Chen, Feiqiong, Qiaoshuang Meng, and Fei Li. 2016. Simulation of Technology Sourcing Overseas Post-Merger Behaviors in a Global Game Model. Journal of Artificial Societies and Social Simulation 19(4) 13: 1.
Cioffi-Revilla, Claudio. 2010. A Methodology for Complex Social Simulations. *Journal of Artificial Societies and Social Simulation* 13 (1): 7.
Cioffi-Revilla, Claudio. 2014. *Introduction to Computational Social Science*. London: Springer.
Crane, Diana. 1972. *Invisible Colleges*. Chicago: University of Chicago Press.
Curry, Timothy J., and Richard M. Emerson. 1970. Balance Theory: A Theory of Interpersonal Attraction? *Sociometry* 33 (2): 216–238.
Davidsson, Paul. 2002. Agent-Based Social Simulation: A Computer Science View. *Journal of Artificial Societies and Social Simulation* 5 (1): 7.
Deffuant, Guillaume. 2006. Comparing Extremism Propagation Patterns in Continuous Opinion Models. *Journal of Artificial Societies and Social Simulation* 9 (3): 8.
Deffuant, Guillaume, Frederic Amblard, and Gérard Weisbuch. 2002. How Can Extremism Prevail? A Study Based on the Relative Agreement Model. *Journal of Artificial Societies and Social Simulation* 5 (4): 1.
Dow, James. 2008. Is Religion an Evolutionary Adaptation? *Journal of Artificial Societies and Social Simulation* 11(2).
Festinger, Leon. 1957. *A Theory of Cognitive Dissonance*. Evanston, Illinois: Row, Peterson.
Friedman, William F. 1918. *An Introduction to Methods for the Solution of Ciphers*. Geneva, Illinois: Riverbank.
Furtado, Bernardo Alves, and Isaque Daniel Rocha Eberhardt. 2016. A Simple Agent-Based Spatial Model of the Economy: Tools for Policy. *Journal of Artificial Societies and Social Simulation* 19 (4): 10.
Gergen, Kenneth J. 1969. *The Psychology of Behavior Exchange, 54*. Reading, Massachusetts: Addison-Wesley.
Gilfillan, S.C. 1963. *The Sociology of Invention*. Cambridge, Massachusetts: MIT Press.
Grosser, Morton. 1962. *The Discovery of Neptune*. Cambridge: Harvard University Press.
Gruber, William H., and Donald G. Marquis. 1969. *Factors in the Transfer of Technology*. Cambridge, Massachusetts: MIT Press.
Hauke, Jonas, Iris Lorscheid, and Matthias Meyer. 2017. Recent Development of Social Simulation as Reflected in JASSS Between 2008 and 2014: A Citation and Co-Citation Analysis. Journal of Artificial Societies and Social Simulation 20(1): 5.
Hegselmann, Rainer, and Ulrich Krause. 2002. Opinion Dynamics and Bounded Confidence: Models, Analysis and Simulation. *Journal of Artificial Societies and Social Simulation* 5 (3): 2.
Heider, Fritz. 1958. *The Psychology of Interpersonal Relations*. New York: Wiley.

Herrnstein, Richard J. 1961. Relative and Absolute Strength of Responses as a Function of Frequency of Reinforcement. *Journal of the Experimental Analysis of Behaviour* 4: 267–272.
Holland, John H. 1975. *Adaptation in Natural and Artificial Systems*. Ann Arbor, Michigan: University of Michigan Press.
Homans, George C. 1974. *Social Behavior: Its Elementary Forms*. New York: Harcourt, Brace and World.
Katz, Elihu. 1960. Communication Research and the Image of Society. *American Journal of Sociology* 65: 435–440.
Katz, Elihu, and Paul F. Lazarsfeld. 1955. *Personal Influence*. Glencoe, Illinois: Free Press.
Kelley, Harold H., John W. Thibaut, Roland Radloff, and David Mundy. 1962. The Development of Cooperation in the 'Minimal Social Situation'. Psychological Monographs 76(19), whole number 538.
Koppes, Clayton R. 1982. *JPL and the American Space Program*. New Haven: Yale University Press.
Kurahashi-Nakamura, Takasumi, Michael Mäs, and Jan Lorenz. 2016. Robust Clustering in Generalized Bounded Confidence Models. *Journal of Artificial Societies and Social Simulation* 19 (4): 7.
McCorduck, Pamela. 2004. *Machines Who Think*. Natick, Massachusetts: A. K. Peters.
Merton, Robert K. 1973. *The Sociology of Science*. Chicago: University of Chicago Press.
Meyer, Matthias, Iris Lorscheid, and Klaus G. Troitzsch. 2009. The Development of Social Simulation as Reflected in the First Ten Years of JASSS: a Citation and Co-Citation Analysis. *Journal of Artificial Societies and Social Simulation* 12 (4): 12.
Minsky, Marvin. 2008. The Emotion Machine: Commonsense Thinking, Artificial Intelligence, and the Future of the Human Mind. The Journal of Personal Cyberconsciousness 3(3)
Minsky, Marvin, and Seymour Papert. 1969. *Perceptrons*. Cambridge, Massachusetts: MIT Press.
Ogburn, William F. 1922. *Social Change*. New York: Huebsch.
Ogburn, William F. 1964. *On Culture and Social Change*. Chicago: University of Chicago Press.
Rogers, Everett M. 1960. *Social Change in Rural Society*. New York: Appleton-Century-Crofts.
Sidowski, Joseph B., L. Benjamin Wyckoff, and Leon Tabory. 1956. The Influence of Reinforcement and Punishment in a Minimal Social Situation. *Journal of Abnormal and Social Psychology* 52: 115–119.
Siedlecki, Patryk, Janusz Szwabinski, and Tomasz Weron. 2016. The Interplay Between Conformity and Anticonformity and Its Polarizing Effect on Society. *Journal of Artificial Societies and Social Simulation* 19 (4): 9.
Smith, John Maynard. 1982. *Evolution and the Theory of Games*. New York: Cambridge University Press.
Stark, Rodney, and William Sims Bainbridge. 1997. *A Theory of Religion*. New York: Toronto/Lang.
Sutherland, Edwin H. 1947. *Principles of Criminology*. Chicago: J. B. Lippincott.
Sutherland, Edwin H., and Donald R. Cressey. 1970. *Principles of Criminology*. New York: Lippincott.
Szocik, Konrad, Kateryna Lysenko-Ryba, Sylwia Banaś, and Sylwia Mazur. 2016. Political and Legal Challenges in a Mars Colony. *Space Policy* 38: 27–29.
Upal, M. Afzal. 2005. Simulating the Emergence of New Religious Movements. *Journal of Artificial Societies and Social Simulation* 8(1)
Urbig, Diemo. 2003. Attitude Dynamics with Limited Verbalisation Capabilities. *Journal of Artificial Societies and Social Simulation* 6 (1): 2.
Urbig, Diemo, Jan Lorenz, and Heiko Herzberg. 2008. Opinion Dynamics: The Effect of the Number of Peers Met at Once. *Journal of Artificial Societies and Social Simulation* 11 (2): 4.

Chapter 3
The Effect of Cultural Drift on Interstellar Colonization

Human civilization is nowhere near the point in its development at which it could establish colonies on the planets of nearby stars, although current technology is probably capable of interstellar communication by means of radio. However, older civilizations that evolved elsewhere in our galaxy may possibly have begun to colonize or at least establish communication channels between them. With the permission of the British Interplanetary Society, this chapter begins with an adapted version of an article I published in its journal, suggesting a sociological theory of interstellar colonization incorporating the concept of *cultural drift*, that evaluates through microcomputer simulation the possibility of numerous limited civilizations, each encompassing a few hundred stars before the wave of expansion stalls (Bainbridge 1983, 1984).

3.1 Beyond the Fermi Paradox

Many serious essays were written decades ago about the paradox that extraterrestrial civilizations capable of interstellar colonization are believed to be numerous across our vast universe, yet none have been observed by Earthlings (Viewing 1975; Cox 1976; Stull 1979; Clarke 1981). This is often called the *Fermi Paradox*, because leading nuclear physicist Enrico Fermi mentioned it in an often-cited 1950 conversation with his colleagues.[1] Many years later, we still lack convincing evidence of the existence of even just one extraterrestrial civilization, and no consensus has consolidated about how to resolve the paradox.

Frank J. Tipler, convinced that unlimited colonization will be undertaken by a significant proportion of advanced civilizations, concluded we may be the only intelligent species in the Galaxy (Tipler 1980, 1981a, 1981b, 1982). His argument is part of a larger challenge to the notion of progress, often called the Anthropic Cosmo-

[1] en.wikipedia.org/wiki/Fermi_paradox, accessed May 2017.

W. S. Bainbridge, *Computer Simulations of Space Societies*, Space and Society,
https://doi.org/10.1007/978-3-319-90560-0_3

logical Principle (Barrow and Tipler 1988). Why is the universe conducive to human life? Because only in such a universe could we evolve and ask the question. In other publications, I have considered the possibility that this thinking implies that the universe was just barely capable of sustaining biological and cultural progress into the twentieth century to a moment I call the *Omicron Point*, when the fundamental question about human existence could be well and fully asked, but perhaps not answered conclusively (Bainbridge 1997, 2017). If the universe really emerged through the equivalent of natural selection from random cosmic variations, just barely sufficient to reach the Omicron Point, then further progress will be impossible. That implies two mutually reinforcing answers to the Fermi Paradox: (1) Extraterrestrials do not exist. (2) Interstellar travel is not feasible.

We do not need a supercomputer to perform a simulation only of these two dichotomies: (1) Extraterrestrial intelligence: Yes or No; (2) Interstellar travel: Feasible or Infeasible. If the true answers are No and Infeasible, there is little to discuss. If the answers are No and Feasible, then the entire cosmos belongs to us. If they are Yes and Infeasible, we may need to shift our simulations to cover interstellar radio communication rather than physical travel. The answers Yes and Feasible stimulate many future questions.

A classic debate occurred between science fiction author Arthur C. Clarke and religious author C. S. Lewis. It took place in print but also in an Oxford pub shortly after the Second World War. Clarke was accompanied by fellow member of the British Interplanetary Society, Val Cleaver (Thomas 1968). Assisting Lewis was his equally technophobic colleague, J. R. R. Tolkien, an expert in historical linguistics best known for his fantasy stories about Hobbits. Lewis had written a trilogy of fantasy novels, which I summarized thus in my statistical study of science fiction authors:

> The first novel of Lewis's trilogy, *Out of the Silent Planet*, says that of the sun's planets only Earth fell from God's grace and thus it must be quarantined so the other planets, still in the Edenic state, will not be contaminated. The villain of the second novel, *Perelandra* is described as a wicked man obsessed with the vile "scientifiction" idea of conquering the universe. The concluding book, *That Hideous Strength*, describes an epic struggle between the scientific forces of evil embodied in a research institute, whose acronym is N.I.C.E., against a few humanistic forces of good embodied in the character of Ransom, a Christ figure, and in the spells of Britain's traditional magician, Merlin (Bainbridge 1986).

In *Perelandra*, Lewis referred to "the vast astronomical distances which are God's quarantine regulations," thereby connecting his theology to the as-yet unnamed Fermi Paradox (Lewis 1996). Thomas Howard has observed that here, "For the hundredth time in Lewis' fiction we are encountering the theme. It is, of course, the old theme of human freedom, choice, obedience, goodness, the will of God, and of how these all harmonize in any sort of pattern (Thomas Howard 1987)." It is worth noting that computer simulations tend to model freedom in terms of so-called *random number generators*, but many of the interstellar simulations described in this book include human actors, thereby opening a window for Lewis' form of freedom to fly in. Despite his enthusiasm for spaceflight, Clarke saw some wisdom in Lewis' perspective, not because it was objective, but because it responded to the currently imperfect development of human culture. A few lines after quoting Lewis's reference to God's quarantine regulations, Clark wrote:

> Little serious thought has been given to the social and moral effects of interplanetary travel, and many people are becoming increasingly critical of those who wish to enlarge man's powers before he is fit to use those he already possesses. The time is coming when - with the lesson of the V-2 behind us - we may have to justify our activities to the world and to prove that the conquest of space will indeed benefit mankind as we have so often proclaimed. If we intend to inflict on other worlds the worst excesses of a materialistic and spiritually barren civilization (and here I am not thinking specifically of modern American culture, but of certain aspects of all Western societies), our case is lost before we begin to plead it (Clarke (1999).

If we think of "God's quarantine regulations" as a metaphor, rather than as a state of faith that God intentionally placed the stars so far apart that only a mature civilization could voyage between them, it can be connected with the Anthropic Principle. For chance to produce one planet in a vast universes on which intelligent life will evolve, there must be many rolls of the cosmic dice - many solar systems and planets within them. Each by chance has a slightly different set of conditions from every other, with perhaps just one being entirely suitable. But the separation of the stars is not limited to the three spatial dimensions, also separating them in terms of the dimension of time. Hundred of millions of times, the suitable planet must circle its star, while the star circles its galaxy. Gradually, the complex chemical system is organized by mutation and natural selection into life then intelligence, through a vast number of molecular events and reproductive generations. Only if the stars exist at great distances from each other, will the frequency of their near collision be so astronomically low that evolution on Earth survives long enough to produce the author and reader of this book.

An alternate possibility, comparable to the Lewis-Clarke quarantine metaphor and suggested at least as early as A. E. van Vogt's 1942 novelette, "Asylum," is that we are currently surrounded by an interstellar civilization which prefers to maintain us in our native state as a zoo or reservation for primitive culture (van Vogt (1942). If so, any attempts on our part to detect extraterrestrial signals are superfluous, since ETs observe us constantly and will make themselves known as soon as we have developed sufficient cultural maturity (Schwartzman 1977). This analysis was popularized in *Star Trek*, as mentioned in the first chapter of this book, in the Prime Directive that advanced civilizations must not influence more primitive civilization until they have autonomously developed interstellar travel. In 2016, João Pedro de Magalhães proposed using radio and television messages to test the zoo hypothesis by demonstrating to any observant extraterrestrials that humanity is ready for direct contact, or at least long-distance communication (de Magalhães 2016).

It is indeed possible that the apparent absence of extraterrestrials in our solar system may result from a tendency of higher civilizations to undertake radio communication in preference to physical exploration and colonization (Tang 1982). Attempts have been made for decades to detect extraterrestrial radio transmissions, without success, so this idea does not resolve the paradox, whatever other merits it may have. In this book, most of the simulations assume physical interplanetary or interstellar travel, with colonization of other planets a frequent scenario. But we shall also entertain the possibility that only information, not human bodies, will travel over cosmic distances. Some of the simulations even apply, with modest reinterpretations, to both scenarios.

In the early years of debate about the Fermi Paradox, most authors ignored the contributions which the social sciences might make toward a resolution of the problem. Both Hart and Singer rejected "sociological explanations" without citing so much as one standard sociological publication, theory or professional school of thought (Hart 1975; Singer 1982). There may exist natural sociological laws which render interstellar societies impossible or which severely limit their expansion. The fact that contemporary social scientists have failed to develop a consistent general model of civilization may reflect the immaturity of the field rather than the impossibility of such an undertaking. This chapter will offer an initial socio-cultural model of interstellar colonization, based in a coherent theory of social behavior, and use computer simulation to show that it is capable of resolving the paradox. Other social theories and computer simulations might also do so, transforming a paradox into a set of competing hypotheses that would require empirical tests, leaving open the question of whether the appropriate research will ever be feasible.

Models of colonization can be based on standard approaches in population biology and population genetics, such as *The Theory of Island Biogeography* by Robert H. MacArthur and Edward O. Wilson (MacArthur and Wilson 1967). That book presents mathematical analyses of species competition in restricted ecological niches as individual members of the species accidentally wander to new environments. While the old theory of panspermia suggests microscopic spores might drift between the stars, interstellar colonization by multicellular organisms cannot happen accidentally, and competition between species will not occur when a single civilization expands into uninhabited regions of the Galaxy. Wilson, among others, extended quantitative evolutionary theory to the study of cultural development, an approach which is more appropriate for the present questions (Lumsden and Wilson 1981; Cavalli-Sforza and Feldman 1981; Bainbridge 1985).

Some classic analyses considered interstellar colonization as a problem in population growth (Jones 1976, 1982; Tinsley 1980). Of course, ordinary colonization cannot proceed faster than the rate of population increase. But other factors, such as those discussed here, may determine lower rates, and colonization is not an effective solution to the problem of population explosion on the home world (von Hoerner 1975). The use of concepts from demography was traditionally quite limited in favor of models drawn, as Newman and Sagan comment, "from blast wave physics, soil science, and, especially, population biology (Newman and Sagan 1981)."

Tipler believes that population biology concepts are appropriate for analyzing colonization by self-reproducing "von Neumann machines," and other authors have suggested that advanced civilizations may produce or evolve into machine cultures capable of unlimited automatic expansion (Clarke 1964; MacGowan and Frederick 1966a, 1966b; Dick 2003). While Tipler's many far-ranging assumptions have drawn intense criticism, I will merely point out two reasons why we should set his radical scenario aside for the time being. First, his ideas seem designed to eliminate all parameters which might limit colonization, while we can best advance our understanding of alternative possibilities through theoretical models of colonization shaped by a variety of restrictive parameters. Second, other writers had not completely succeeded in showing how the paradox of an uncolonized Earth could be resolved in any way other than Tipler's, a task whose accomplishment is sketched below.

3.2 A Theoretical Background

The previous chapter considered how a colony on a nearby planet might organize itself, but did not explain how social forces here on Earth would motivate the great expense of sending sufficient people and resources across millions of miles of empty space. To consider all the social science theories that might inform the challenges and opportunities for interplanetary colonization would require far more than just one book, so it is reasonable to focus on a particular subset that I have found exceedingly appropriate for computer simulation.

Of all the competing schools of sociological thought, the one closest in approach to the physical sciences is sometimes called Behaviorism. Disliking "isms" and convinced that parallel developments in cognitive science, economics, psychology, anthropology, ethology and sociology should be combined, I prefer the term *Behavioral Science*. This theory uses not the gene or the reproducing population as its unit of analysis, but the reward-seeking individual animal or person. It seeks to derive the behavior of all organisms, including all possible species of intelligent, social beings, from abstract "axioms" presumed to be universal in application, although discovered from experiment and observation of terrestrial rats, pigeons and persons. In his monumental *Sociobiology, the New Synthesis*, Edward O. Wilson said this approach is compatible with theories of behavior based in evolutionary biology, so it is quite possible that it represents the intellectual seeds from which a definitive science of society will grow, despite the fact that most sociologists currently work from other perspectives (Wilson 1975).

In 1938, B. F. Skinner stated the key proposition: "If the occurrence of an operant [unit of spontaneous behavior] is followed by presentation of a reinforcing stimulus, the strength is increased (Skinner 1938)." Applied to human beings, this became the first of five propositions from which George C. Homans believed he could derive all major forms of social behavior: "For all actions taken by persons, the more often a particular action of a person is rewarded, the more likely the person is to perform that action (Homans 1974)." Upon such simple and familiar foundations, some rather elaborate theoretical structures have been built. Already in 1941, Neil E. Miller and John Dollard showed that even the subtle tendency of humans to imitate the behavior of others could be explained through such ideas (Miller and Dollard 1941). The same approach has been used by Peter M. Blau to analyze social power which some individuals may gain over others (Blau 1964). The previous chapter reports research building simulations of religious faith and social movements on these principles, so it is possible to derive theories of superficially non-pragmatic social behavior, but evoking the uncertainties of chaos and complexity in large systems of interaction.

This theoretical tradition suggests two main propositions with implications for interstellar colonization. First, large social structures (such as civilizations) should be precarious constructions, liable to collapse because they are so far removed from the social exchanges between individuals which are postulated to be the basic constituents of society (Homans 1950, 1967, 1984). Competing social theories, such as Marxism, postulate the existence of regular laws operating on the level of whole societies, thus capable of providing permanent societal coherence and guiding evolution to higher levels of development such as interstellar civilization. But if Skinner and Homans are correct, such laws do not exist, and all social behavior ultimately serves individual needs.

Second, classical Behavioral Science predicts that societies will not undertake collective projects unless these give a high, short-term profit to those individuals who invest in them. Together, these propositions seem to preclude interstellar colonization, because the first suggests civilizations will collapse before they colonize, while the second says they will have no motive to plant commercially unprofitable colonies at huge expense across the vast gulfs of interstellar space. The concluding chapter of this book will mention a very different set of theories, those concerning the rise and fall of civilizations. They are not currently popular within social science, but neither is spaceflight.

However, the application of Behavioral Science principles to the understanding of religious commitment and radical social movements, introduced in the previous chapter, has shown that a wide range of historical developments may sometimes occur even if they appear quite irrational in terms of individual material gain. From this perspective, in 1976, I wrote that spaceflight and colonization were unlikely undertakings for a civilization, but not impossible (Bainbridge 1976). I argued that they could happen under unstable social conditions with the cooperation of historical accidents, although under "normal" conditions in societies they would not take place. In the proceedings of the Third History Symposium of the American Astronautical Society, back in 1982, I wrote:

> I think it is very likely that ancient intelligent species will have evolved static societies which have achieved containment of instinct - which have transformed themselves culturally and biologically to satisfy natural desires in the most efficient and safest ways. The social conditions which magnify aggressive and exploratory drives are highly dangerous. They generate pressures toward war before they generate pressures toward interstellar colonization. Perhaps such "outward urges" could fuel an ambitious space program. But for individuals and small groups, quicker rewards of power and prosperity can be obtained through competition against other individuals and groups. Thus, all societies which continue to be aggressive and expansionist will be politically unstable. When they reach our level of technological development, they enter a period of extreme danger, and the risk of nuclear annihilation is only the most obvious of the ways they could bring doom upon themselves (Bainbridge 1982).

These assumptions may or may not be correct, but they are plausible and can be derived from at least one contemporary social-scientific theory. In what follows, my task is to examine through computer modelling what the implications are for the rate and shape of interstellar expansion. While our approach will be similar to that of work based on population biology, the key concept will concern not biology but culture.

3.3 Cultural Drift

In the Behavioral Science model, there will be rare times when a civilization develops political tensions, social movements, a conquering religion, or some other unstable structure capable of mobilizing great social energies for the accomplishment of a project lacking ordinary utilitarian justification. Since my first book was a social history of the development of modern space rocketry written from this perspective, I naturally imagine scenarios of interstellar colonization involving a spaceflight social movement able to exploit dangerous political conflict such as that between the US, USSR and Nazi Germany. But other scenarios are possible. All we assume is that a civilization somehow develops a culture with an unusually high probability of colonizing (Molton 1978; Bainbridge 1991, 2015).

Societies change, often at random in areas which are not crucial to the survival or economic prosperity of their members. Random cultural change, likely in areas which are not of great practical value to members of the society, is called *cultural drift*. Perhaps the most obvious example is in language. Over time, pronunciation, spelling, vocabulary and even grammar will change. While some of this is evolution in response to changes in technology and natural conditions, much of it apparently represents a "drunkard's walk" chance wandering of the language among an infinite number of equally-functional alternatives. Matthew W. Hahn and Alexander Bentley have applied the concept of cultural drift to the example of the names parents give to their babies, usually selecting from an existing set of names, but with the equivalent of statistical sampling errors that gradually change the content of that set of names (Hahn and Bentley 2003).

While this random process will produce even unlikely outcomes, so long as they are possible, these outcomes will typically be rare not only because they are hard to find but also because they are easily lost. A familiar statistical phenomenon known as "regression toward the mean" represents the tendency of a case (here a society) with an extreme value on some variable (here the probability of colonizing) to change in the direction of the typical value of the variable. That is, a culture which does beat the odds and begins to colonize will tend to drift back away from colonization.

How can we represent a cultural disposition to colonize and cultural drift away from this characteristic? I have already presented the simple idea of probability of colonizing within an historical era. How long an "era" is in years matters not very much for present purposes, because a low probability in a single year can be represented by a higher probability in a century and a still higher probability in a millennium. Here we will let "era" be some significant period of time, comparable to that required for a colony to be established and grow to the point that it would be capable of sending out colonies of its own. The probability of actually colonizing in such a period we represent by C, which, since it is a probability, is a real number between zero and one.

At this stage in our theorizing, we cannot give a definite law for cultural drift. However, for the purposes of simulation, we can suggest an abstract approximation of what such a law should look like. We can imagine a series of values for C that change over time: $C_0, C_1, C_2, C_3 \ldots C_N$. We want a formula which can be applied to C to produce a new value for the probability of colonizing, that is, one which will generate C_1 as a function of C_0. Obviously, C_1, and all possible descendants of it resulting from repeated application of the formula, must, like C_0, range between zero and one. The simplest function which accomplishes this lets C_1 equal C_0 taken to some positive exponent, X, such that $C_1 = C_0^X$.

What factors should we put into X? Since we want to use the formula in computer simulations, we should have some parameter, P, which will be a constant in any given run but can be varied across different simulations and which can be used to fine-tune the program. Also, a key concept in this approach is that chance plays a crucial role in cultural drift. Therefore, X should also include a random factor, R. As it happens, the BASIC language in which these simulations were written back in 1983, like many others, includes a function which produces essentially random numbers ranging between zero and one. The newer program used in the final section of this chapter was written in the Pascal language; I have used the C language for other programs, and both have comparable randomization procedures. Consideration of a few alternatives convinced me that the best way to combine P and R into X was simply to add them $(X = P + R)$. Or, to state the cultural drift transformation function in BASIC:

$$C1 = C0^\wedge(P + RND(1))$$

This formula incorporates the assumptions we have drawn from Behavioral Science theory in the simplest way which will work in a computer simulation. Of course, one can imagine quite other formulae, and sociology is not yet able to specify the version providing the closest approximation to reality. In demography, as Nathan Keyfitz has pointed out, a simple formula often provides the best approximation to reality, although one cannot be sure, of course, in the absence of data (Keyfitz 1982). The reader will see how the formula actually operates below, and the original article in the *Journal of the British Interplanetary Society* provided the full BASIC program so that interested readers can experiment with alternative formulae, as well as with different values for the parameters and extensions of the model to include other ideas as well. Table 3.1 shows how the formula operates with a variety of values for the parameter P.

Table 3.1 does not report results of full simulations, but instead gives the distribution of outcomes after applying the formula 5 and 10 times to an initial probability $C = 0.5$ of colonizing in a given era. We imagine that a home world with initial $C = 0.5$ plants a colony at another star. In doing so it also plants its culture, but the culture drifts during the colonization process. Exactly which technical means are used to establish the colony is not crucial. Heinlein has suggested that cultural drift

Table 3.1 Distribution of drift effects over 1,000 trials for each parameter value

P (Drift formula parameter)	After 5 generations		After 10 generations	
	Mean C	Percent 0.5+	Mean C	Percent 0.5+
0.50	0.547	61.2	0.592	65.5
0.55	0.463	46.4	0.446	45.4
0.60	0.375	30.2	0.286	23.6
0.65	0.299	17.9	0.169	9.3
0.70	0.213	6.2	0.075	1.8
0.75	0.149	3.1	0.028	0.6
0.80	0.096	0.5	0.007	0.0
0.61803	0.350	25.3	0.248	16.5

is quite likely in city-sized spaceships which take generations to reach their destinations, while Hogan has said the same will happen even in small, fast ships which carry human genetic material from which to build a colony automatically (Heinlein 1941; Hogan 1982). So when a colony is planted we apply the formula to its parent's culture to determine the new C_1 of the first generation colony. After a suitable time, the colony may itself colonize, and we apply the formula to C_1 which gives us C_2. From C_2 we get C_3, and so on to C_5 and C_{10} which are reported in Table 3.1. Each trial represents a chain of colonization linking many worlds, the home world and ten generations of offspring.

Since our formula includes R, the random factor, each time we calculate out 5 or 10 generations, we will get a different result. Therefore, Table 3.1 gives the results not of single runs but rather the distribution of outcomes from 1,000 trials for each value of P. In the first row, we look at $P=0.5$, for example. The mean value of C_5 after 1,000 trials is 0.547, and after 10 generations of colonies, the average C_{10} is 0.592. Thus, for $P=0.5$ the value of C gradually rises, on average. In 61.2 per cent of the trials $C_5 > 0.5$, while in 65.5 per cent, $C_{10} > 0.5$. Thus, in a majority of cases, the probability of colonizing increases gradually, while in a smaller number of cases, the probability decreases.

But our theory indicates that colonization is generally unlikely, and the typical C should be low. Therefore, cultural drift should make C drop most of the time, which is the case for the other rows in the table. For $P=0.8$, the rate of drop is very steep, and not a single one of the 1,000 trials at that value for P results in $C_{10} > 0.5$.

There is a reason for preferring the value $P=0.618$ (or $P=0.618034\ldots$), a justification which has some theoretical force, albeit weak. This happens to be the positive number which differs from its reciprocal by one, the so-called *Golden Ratio* that has been invested with philosophical meaning since the ancient Greeks, and will appear again in Chap. 7.[2] Since R ranges between zero and one, this means that X, which is $P+R$, ranges from 0.618 to 1.618. Since this establishes the range of the exponent,

[2] en.wikipedia.org/wiki/Golden_ratio, accessed December 2017.

X, and $0.618 = 1/1.618$, then regardless of what happens in one step of cultural drift, a second step can exactly reverse it. At $P = 0.618$, there is a very moderate cultural drift downward in C, entirely caused by the fact that 61.8 per cent of the time X will be greater than 1, and if $0 < C < 1$ and $X > 1$ then C^X must be less than C. One could consider other values for P, but 0.618 is a very gentle assumption of downward drift to begin with.

3.4 Colonization in a Sphere of 1,000 Stars

The simulations reported here all track the expansion of a single civilization with its home world at the center of a sphere of 1,000 colonizable solar systems. What counts as "colonizable" depends both on the technology available to the colonizers and on the social motivations which impel the expansion. If moons like Rhea and Iapetus are colonizable, than many solar systems are open to colonies, since these satellites of Saturn must represent a very common type of body. If, on the other hand, very Earthlike planets are required, colonizable solar systems must be, on average, further apart (Bond and Martin 1978; Bond et al. 1980; Lissauer 2000).[3] With the home world at the center, colonizable stars are distributed randomly in the volume of a sphere 1,000 distance units across. The size of this unit takes on meaning only in relation to the maximum range of colonizing spaceships, and this range will be a parameter varied across simulations (Walters et al. 1980). Therefore it is not necessary to decide the absolute diameter of the sphere, although for those who want a definite image of our star globe, I suggest a diameter between 100 light years and 100 parsecs.

In the first simulation, we will let the maximum flight range of spaceships be 100 units, begin with a home world which colonizes with a cultural probability of doing so equal to one half ($C = 0.5$), and set the cultural drift coefficient equal to 0.618. The run goes for 15 eras, and in each one every inhabited solar system which has not already colonized will be given a chance to do so equal to its value of C. When a star colonizes, it plants offspring on every uncolonized world within the reach of its ships, here a radius of 100 units. Each new colony gets a value of C derived by applying the formula to the C of its parent. If a star misses the opportunity to colonize in a given era, cultural drift still occurs, but through a change in its own culture. That is, the formula is applied to the non-colonizing star's old C, to produce a new C value which will be its chance of colonizing in the following era. While different runs will give different results, the one reported in Table 3.2 is quite typical.

There are nine stars in the first generation of colonies, so counting the home world there are ten societies after the first era. The farthest colony is 93.4 distance units from the home world, while the nearest uncolonized star, just beyond the 100-unit range of the ships, is 103.8 units away. In the second generation, 12 societies are added,

[3]ntrs.nasa.gov/search.jsp?R=20000115621&hterms=Conservative+Estimate+Number+ Habitable+Planets+Galaxy

Table 3.2 Simulation I: colonization with cultural drift

Generation	Societies	Distance from home world to:	
		Nearest system not colonized	Farthest system colonized
1	10	103.8	93.4
2	22	114.0	146.3
3	33	118.4	222.1
4	63	166.7	299.3
5	99	168.9	389.2
6	167	170.5	451.2
7	251	170.5	476.4
8	374	247.9	495.9
9	531	248.6	495.9
10	674	252.9	499.3
11	833	290.7	499.4
12	916	290.7	499.4
13	948	290.7	499.4
14	965	290.7	499.4
15	975	389.6	499.4

bringing the total to 22. The shape of the interstellar civilization has begun to get lopsided, since the farthest colonized star at 146.3 units is now more distant from the home world than the nearest star not colonized, at 114.0. By the fifth generation, when the total number of societies has risen to 99, this irregularity is more pronounced, the farthest colony being 2.3 times the distance of the nearest uncolonized star, 389.2 units compared with 168.9. In the next generation the expanding wave-front of colonization comes within reach of the surface of the sphere of stars, and in the eleventh achieves its maximum distance from the home world, 499.4 units.

Two facts stand out in Table 3.2. First, as just noted, the expansion becomes rather irregular in form, with some near-in stars escaping colonization until quite late (Jones 1978; Smith 1981). Second, the expansion takes rather longer than it might without cultural drift. A star only 290.7 units distant from the home world is not colonized until the 15th generation, although ships might have reached it in the third or fourth generation if distance were the only factor. The range of the ships, 100 units, would take them to the surface of the sphere in only five flights, but of course the colonization process must use stars as stepping stones, and they are distributed randomly. So part of the reduction in speed of the colonization wave is due to star placement and part, only, to cultural drift.

The fact that our computer simulation includes 1,000 stars, and not a much larger number, means that the ships will reach the edge soon in some direction and be unable to continue colonization beyond the radius of 500 units from the home star. Of course, to double this radius with the same density of stars would require increasing the total number to 8,000, and increasing the radius by a factor of ten would require a simulation with 1,000,000 stars. The original program searches through the entire list of uncolonized stars each time a society colonizes, to see which are within the flight distance, a time-consuming procedure. Although this time could be reduced through various programming tricks in cases where the maximum flight distance was a small proportion of the sphere radius, in general the time the program runs is an exponential function of the number of stars. The concluding section of this chapter will employ a much more recent program written for considerably more powerful personal computers than was available back in 1983, representing a larger number of stars.

Even if our sphere contained one million stars, the tardy society 290.7 units out (not established until generation 15 in Simulation I) could not have been planted much earlier. Imagine a salient or pseudopod of colonization, carrying culture with a high C out in some direction from the home star, and a tardy section of wave front a few tens of degrees of arc away from it. While the more active salient could reach out and then send an arm of colonization back into this hole, it would take it a few generations to do so, even if no colonization opportunities were missed. Of course, in the real galaxy there are rifts in space, like gaps between spiral arms and areas around gaseous nebulae, where the density of habitable planets is unusually low. We should be aware of the limitation imposed on our simulations by the surface of the star sphere, but can discover the chief principles illuminated by our model quite effectively within it.

3.5 Effect of Spaceship Range

The wave of colonization appears to slow down in Table 3.2, adding the largest number of societies in generations 9, 10 and 11, trailing off after that. Of course, it has reached the edge of the sphere and approached the end of the supply of uncolonized stars at this point. But if the wave moved more slowly, perhaps it would die out well within the 1,000 star limit. Simulation II, described in Table 3.3, explores this possibility by setting a more modest ship range, 80 units rather than 100. This slight difference in range produces a large difference in the number of stars which can be reached from any starting point, however, on average reducing it by almost exactly half. While the farthest edge of the interstellar civilization reaches 490 units in the 8th generation of Simulation I, in Simulation II it takes 12 generations. At this pace, one would project that the second simulation would achieve before 25 generations what the first did in 15, but Table 3 shows that this is far from the case.

After 25 generations, this simulation with a ship range of 80 units has colonized only slightly over a third of the stars, compared with nearly all after 15 generations in the first run. The nearest uncolonized star is only 159.9 units from the home world, ideally just two flights out. It may represent our own solar system, and potentially resolves the Fermi paradox: Bad luck has placed us too far from the nearest extraterrestrial civilization. The rate of colonization reaches its maximum between generations 11 and 15, and drops by a third for the last ten generations. While the

Table 3.3 Simulation II: reduced maximum flight

Generation	Societies	Distance from home world to:	
		Nearest system not colonized	Farthest system colonized
1	7	80.0	77.8
2	15	93.4	125.0
3	16	93.4	166.7
4	25	93.4	241.8
5	32	93.4	253.7
6	41	102.3	278.8
7	48	102.3	302.1
8	59	102.3	338.2
9	67	102.3	408.4
10	77	102.3	445.9
11	102	102.3	486.0
12	125	118.4	490.3
13	163	118.4	490.3
14	186	159.9	490.3
15	210	159.9	492.5
16	229	159.9	492.5
17	239	159.9	492.5
18	263	159.9	499.9
19	276	159.9	499.9
20	294	159.9	499.9
21	318	159.9	499.9
22	334	159.9	499.9
23	347	159.9	499.9
24	364	159.9	499.9
25	377	159.9	499.9

simulation ends before colonization has ceased, clearly it will take a very long time to incorporate the whole star cluster in the slowly expanding civilization.

Because chance plays a major role in these simulations, both in the operation of cultural drift and in the original placement of stars, runs using exactly the same parameters can come to very different results. A second attempt identical in assumptions to Simulation II resulted in 92 societies at generation 25, rather than 377. It stalled completely after planting the 166th colony in generation 68, with a probability of colonizing onward (C) of only 0.009. In this case, cultural drift brought expansion completely to a halt. Of course a ship range of only 80 units may mean that many stars cannot be reached at all, and it is worth trying runs in which C is always 1 and drift downward does not occur, to determine the maximum likely rate of expansion. In one such, ships with a range of 80 units built a civilization of 292 societies after 10 generations, nearly half the rate in Simulation I and about four times that in Simulation II.

3.6 Effect of Initial Colonization Probability

Rather than varying ship range, we can adjust the initial probability of colonization, C. A high initial C represents a rare culture in which colonization is highly likely. A low initial C represents a society in which very special political or other historical events of low probability might start colonization, but in which the culture transmitted to colonies is not especially conducive to further colonization. Thus, one might think that we can stall colonization simply by lowering C. All the runs reported in this chapter assume the home world will colonize at its first opportunity, rather than missing early chances and letting its C drift downward and downward. Of course, one could set C abysmally low, so that a second generation of colonies would be almost impossible. But I think the most elegant simulations would permit development of a multi-generation interstellar civilization before expansion stalls, so we will inspect what happens if we lower C only moderately below the 0.5 level.

Table 3.4 reports results of Simulation III in which the ship range was set back to 100 units but initial C was reduced from 0.5 to 0.4. After 10 generations, the expansion has fallen behind that of Simulation I, producing a civilization of 321 societies compared with 674. But by generation 15 it has begun to catch up, and at generation 40 achieves what the first simulation did by generation 15, a total of 975 societies. At this point, Simulation III stalls, and even by generation 75, no further stars have been colonized. While 25 stars remained uncolonized, clearly something happened around generation 10 that kept expansion from stalling before a majority of stars were colonized, a factor capable of overcoming the disadvantage of a low initial probability of colonizing.

A principle of natural selection is responsible (Hanson 1998). Recall that cultural drift can increase C as well as reduce it, although on average the majority of colonies will have lower Cs than the societies which established them. But offspring with increased C will have more chance of colonizing immediately than their siblings with lower C. If a low-C society fails to colonize in one generation, its C will tend to drift further down, and when chance does let it colonize, all worlds within a 100-unit radius may already be inhabited. Colonizing societies with higher than average C will transmit this superiority to their progeny. While cultural drift will tend to reduce C across generations, natural selection will tend to increase it.

We can compare the actual behavior of Simulation III with results from a run like those reported in Table 3.1. On 1,000 trials, starting with C = 0.4, after ten generations of iterating the drift formula the mean C dropped to 0.177, and in (coincidentally) only 17.7 per cent of the trials did C end up greater than the initial value of 0.4. On the tenth generation of Simulation III, 65 new societies were founded. Their mean C was 0.324, much higher than the 1,000 trials predicted, and 32 per cent of them had Cs greater than 0.4. Moreover, values of C tended to move upward after this,

Table 3.4 Simulation III: reduced initial colonization probability

Generation	Societies	Distance from home world to:	
		Nearest system not colonized	Farthest system colonized
1	6	102.4	98.7
2	12	103.8	150.5
3	20	103.8	204.3
4	35	103.8	250.2
5	58	115.4	295.2
6	103	130.4	348.7
7	156	130.4	358.1
8	196	130.4	432.3
9	256	130.4	488.8
10	321	130.4	496.6
11	403	130.4	497.4
12	499	148.6	497.9
13	584	183.1	499.3
14	687	245.3	499.3
15	761	298.4	499.6
20	921	298.4	499.6
25	951	298.4	499.6
30	960	298.4	499.6
35	974	298.4	499.6
40	975	298.4	499.6
75	975	298.4	499.6

rather than drifting down. On the 15th generation, 74 colonies were established, with a mean C of 0.459, 66 per cent of them having higher probabilities of colonizing than the initial 0.4. Thus natural selection overcame downward cultural drift under these conditions.

Obviously, one factor influencing the outcome is how fiercely the formula tends to drive C down. On the other side, a factor encouraging upward movement from natural selection is the number of offspring produced by the typical high-C colonizing society. Since such societies will tend to try colonization before their low-C sibling neighbors, a key determinant of this is simply the range of the space ships - or, more precisely, the typical number of uncolonized stars within the range of the ships. Even if the pressure for downward cultural drift is strong, a large number of progeny in each generation means a good chance that at least one will have higher C than its parent and be able to transmit this advantage to at least some of its own offspring.

I did a postmortem analysis of Simulation III to determine why the 25 holdout stars never got colonized, even given 75 generations. A special program took the positions of the 1,000 stars and determined that only three could not be reached by 100-unit flights from the home-world, two of them in a pair 70 units apart. The remaining 22, which would have been colonized but for cultural drift, included two isolated stars and a pair near the surface of the sphere, as well as the uncolonized star only 298.4 units from the center. Finally, there were two uncolonized pockets, about 600 units apart, each reaching in from the surface more than 100 units, one with 6 stars and one with 10.

The star which remained uncolonized, yet was only 298.4 units from the home star, is especially interesting. Since it is 200 units inside the sphere, it could not be colonized from outside unless it could be reached from another uncolonized star, but it was fully 332 units from the nearest of them. However, it was only 88 units from the 163rd colony established in the run, and thus could have been reached from the home world. If the Sun were that lone star, we could be living deep within a huge interstellar empire, untouched by an explosion of colonization which swept around us many years ago.

It is worth tracing the steps by which the civilization could have crossed the 298.4 distance units to the lone star, to get a more vivid picture of the process. Colony number 2 was established in the first generation with a C of 0.246, drifted downward from the 0.4 possessed by the home world. In the second generation, only its sibling, Colony 3, was able to plant colonies, and Colony 2s C drifted down from 0.246 to 0.162. In generation 3, the drift was upward to 0.320, while in generation 4 it was down again to 0.228. Finally, in generation 5, Colony 2 was able to produce two offspring, among them Colony 41 with 0.112 for its C. In generation 6, Colony 41 missed a chance to colonize, but drift brought its C up to 0.236. In generation 7, Colony 41 produced four offspring, including Colony 112 with C = 0.301, and in generation 8 Colony 112 had a single offspring, Colony 163 with C = 0.402.

This colony, beginning with a relatively favorable probability of having offspring, was the only world from which the wave of colonization could reach the lone star. But if it didn't colonize quickly, its C was likely to drift to a prohibitively low level. Generation 9 saw Colony 163s C drop to 0.245. It dropped further to 0.160 in generation 10 and wandered, sometimes rising but mainly dropping until it reached 0.001 in generation 18. In generation 19 it drifted below 0.0005 and was rounded off to zero. Thus it lost its chance to colonize the lone star. But for this to happen on a wide front, the rate of downward drift would need to be much higher, capable of offsetting natural selection.

3.7 The Power of Drift

Table 3.1 gives us one way to achieve more rapid cultural drift, increasing P above the level of 0.618 we have been using. Partly because 0.618 affords theoretically satisfying symmetry, and partly to show the range of options offered by the simulation approach, I shall use a different method, revision of the drift formula. Since R is a random number between zero and one, we can control the power of cultural drift by changing the distribution of random numbers within this range. The simplest way is to take the square root of R. Since we are dealing in abstract models, and sociology cannot yet specify the form or terms of the correct formula, I cannot offer a strong defense for this move. However, it is a simple change which achieves the desired end without disturbing the range over which C can drift. In BASIC, the formula thus becomes:

$$C1 = C0\hat{}(P + SQR(RND(1)))$$

This revised formula makes C decline rapidly. In a test of 2,000 trials, similar to those in Table 1 but run twice as many times, the mean value of C after five generations was 0.091, and after ten generations was 0.006. In about half a per cent of the cases C rose in the first five generations, but never wound up greater than the initial 0.5 after ten generations. It might seem unfair to the colonizing societies to shift to such a harsh rate of drift after the easy-going demands of the formula we have been using. But I must emphasize two things. First, if there exist actual rigid laws of social evolution, applicable to all advanced technical species, they may imply rapid downward drift in colonization, or they may not, but the suspicion of Behavioral Science that large-scale social phenomena are highly unstable suggests rapid drift. Second, the new formula does not really demand very rapid cultural change. The flight time of an interstellar fleet could well be several hundred years, as could the time required to settle a new world and think about colonizing onward, so ten generations could well consume several thousand years.

Table 3.5 shows the results of our last run, Simulation IV, in which the initial C is 0.5, the ship range is 100, and we use the new formula for drift. The expansion starts out rather vigorously, but stops altogether on the 17th generation, with just a third of the stars colonized and the nearest uninhabited world a mere 139.2 units from the home star, potentially accessible in only two flights. The last star with any chance to plant colonies lost it in generation 26 when its C drifted from 0.002 to zero. The results of this simulation are extremely interesting. If it should happen to model the real universe at all closely, then there may exist many interstellar civilizations, each encompassing a few hundred star systems, yet none spreading out to fill the entire galaxy.

3.8 Principles of Drift-Influenced Interstellar Colonization

If the general approach of this chapter is correct, given values for our parameters which fall in certain wide ranges, then the following propositions could be true:

1. Interstellar civilizations are highly irregular in shape.
2. Uncolonized habitable worlds exist near the home worlds of large interstellar civilizations.
3. The expansion of interstellar civilizations proceeds more slowly than expected by models which do not incorporate the concept of cultural drift.

Table 3.5 Simulation IV: alternative drift algorithm

Generation	Societies	Distance from home world to:	
		Nearest system not colonized	Farthest system colonized
1	9	103.0	98.1
2	21	106.1	171.8
3	36	106.1	259.4
4	67	106.1	350.4
5	102	113.1	359.4
6	164	113.1	439.9
7	212	113.1	449.4
8	246	139.2	498.7
9	276	139.2	498.7
10	293	139.2	499.8
11	312	139.2	499.8
12	323	139.2	499.8
13	326	139.2	499.8
14	328	139.2	499.8
15	330	139.2	499.8
16	331	139.2	499.8
17	332	139.2	499.8
75	332	139.2	499.8

4. The expansion of interstellar civilizations slows nearly to a halt after a moderate number of worlds have been colonized.
5. A single intelligent species will not quickly colonize an entire galaxy.
6. The observation that the Earth has not been touched by an interstellar civilization does not imply the non-existence of such civilizations.
7. Models of interstellar colonization employing the concept of cultural drift strengthen arguments for the attempt to detect extraterrestrial radio signals.

If the premises drawn from Behavioral Science theory are incorrect, some of these propositions are invalid. For some sets of parameters, natural selection overcomes cultural drift, even within the theoretical model offered here. We have assumed that colonization is unlikely, that some societies nonetheless will develop cultures making colonization more likely, and that cultural drift will tend to bring these elevated probabilities back down near zero. The fact that we were able to model limited colonization renders these assumptions especially attractive, since we can now postulate the existence of nearby interstellar societies despite the apparent absence of extraterrestrials on Earth.

How can computer simulation methods be applied in interstellar communication via radio? Most obviously, we could reinterpret the original 1983 simulations as involving 1,000 indigenous extraterrestrial civilizations surrounding the home solar system. Thus the action would not be sending colony fleets from star to star, but messages that could be decoded (Cameron 1963; MacGowan and Frederick 1966a, 1966b; Sklovskii and Sagan 1966; Swift 1990). In the 2011 book *Civilizations Beyond Earth*, edited by Douglas Vakoch and Albert Harrison, I had explored the possibility that computer simulations of individual human beings could be transmitted to a receptive alien civilization, thereby allowing a kind of virtual direct contact, literally at the speed of light (Bainbridge 2011). Here we need not assume the technical possibility of this alternative form of personal interstellar travel, but merely accept the much more reasonable premise that radio communication across the gulfs of space to other civilizations is possible.

One potential consequence would be the sharing of cultural traits, which most obviously could begin with scientific data, such as two civilizations comparing their astronomical observations and learning from the differences in their genetic codes. Yet an interesting possibility is that religious faiths might migrate via interstellar radio communication. It is worth noting that major religions of today's Earth are popular far from their geographic points of origin, Buddhism having migrated from India to Japan, and Christianity from Israel to Argentina. In the previous chapter, we modelled religious behavior on Mars, and here we can apply similar principles to a wider universe.

The simulations featured in the previous chapter were multi-agent systems, but the word *agent* is open to several meanings. Most obviously, an agent is a computer simulation of an individual person. But agents represent any entity that interacts with its environment and makes decisions about which actions to take. Thus an agent can represent a civilization, so long as the civilizations to be modeled have relatively coherent mechanisms for collective decision making.

The models of religion reported in the previous chapter were repeatedly updated as readily available computers improved, reported for example in a 1995 journal article, "Neural Network Models of Religious Belief," and a 2006 book, *God from the Machine: Artificial Intelligence Models of Religious Cognition* (Bainbridge 1995, 2006). Here we shall use a current version of the software derived from the 1987 model of interpersonal influence across a social network, but conceptualize the agents not as individual people but as separate intelligent species living in different solar systems. Rather than model a sphere of stars located at random, the simulation is more abstract, arranging the agents in a way similar to the last example in the previous chapter, across a two-dimensional *flatland* (Abbott 1884). Instead of using spaceship range as one of the defining parameters, here agents could communicate only with the eight surrounding agents.

The new matrix is just as abstract and two-dimensional as the original 1987 one with 190 agents, but much larger, consisting of 44,100 agents arranged in a square 210 on a side. The reason for selecting these numbers is that 44,100 is evenly divisible by 2 (giving 22,050), 3 (=14,700), 4 (=11,025), 5 (=8,820), 6 (=7,350), 7 (=6,300), 9 (=4,900), and 10 (=4,410), leaving out only 8 which would have required four times the population, to have a perfect square into which all 10 of the lowest integers could divide evenly. The reason why such numbers are desirable for an abstract simulation such as this rather complex one, is that the population can be divided into exactly equal subgroups in many ways. The controls include many more options than in the original program, and of course the program code can be revised to support a particular study.

Here is one run of the simulation, which we can imagine represents a galaxy having 44,100 independent civilizations. In this particular case, we postulate that 6 distinct types of religion might exist, for example: (1) monotheism with a savior, (2) monotheism without a savior, (3) polytheism, (4) spiritual withdrawal through meditation, (5) pseudoscience like a combination of psychoanalysis with parapsychology, and (6) atheism. The simulation does not attempt to model these different forms of cognition about the meaning of life, but merely marks each civilization's culture as belonging to one category or another. If two civilizations belonging to the same religious category come into radio contact with each other, they may share minor cultural symbols, just as the Greeks and Romans equated Zeus with Jupiter, but will primarily reinforce the existing faith that each possesses. If two civilizations belonging to different categories come into contact, they may try to convert each other, as has happened many times when Christian cultures tried to assimilate Jewish people.

The particular run of the simulation reported here began with six distinct religious categories, intended to be of equal size. But the selected setting of the simulation did not force exactly 1/6 or 7,350 civilizations into each category, but gave each a 1/6 chance of belonging to any particular category, which means that the software's random number generator produced a range of numbers, from 7,215 to 7,451. Another run would produce a different pattern, but if social influence is a factor, this gives a lucky advantage to the largest group.

Figure 3.1 shows what the very center of the galaxy looks like, representing each civilization by a tiny square colored to represent its category, with a social network density of 50%, which was but one among many choices decided before beginning the simulation. The probability that balance theory operating on social bonds would be applied, rather than differential association theory operating on category membership, was also set at 50%. Four of the categories were set so that a member's chance of developing an enduring communication link with a neighboring agent in a different category, thus engaging in outreach, was set at 0%. For group number 1, in contrast, the outreach probability on a given turn was set at 60%, and for group 2, at 40%. We might imagine that these are the two forms of monotheism, the one with a savior having more evangelical fervor than the one lacking belief in a messiah. Thus we are modeling a religious marketplace in which only two of the six categories are actively recruiting, and at different levels of intensity. We do not actually need to define which ideology each group has, beyond this measure of propensity to evangelize.

The image shows just 400 civilizations in the galaxy, less than 10%, and the software allows the user to scan across the galaxy, slowly or rapidly. Thus, I could have created a total map, but reproducing it here would have rendered the agents and bonds practically invisible, one more indication of how modern computers are much more powerful than decades-old machines. I also could have grouped all members of one category in a central location, but chose not to do that for this illustrative

Fig. 3.1 The center of a galaxy of 44,100 civilizations, initial conditions

simulation. Table 3.6 shows population statistics at the beginning and over the first 100 turns of interaction.

As luck would have it, Group 1 started with slightly greater population as well as being aggressive in recruiting. It increases by 4,325 members over the first 10 turns, a spectacular growth rate of 58%. We can imagine that these turns take decades or even centuries, not merely because of the round-trip time for interstellar radio signals to aggregate into a full conversation, but also for a civilization to transform its social institutions and culture to conform to a new religious orientation.

Group 2 was unlucky to start with the lowest population, but with its more modest recruitment efforts was able to grow by 984 or 14%. All the other groups lost membership. The final column of the table lists the total number of conversions to that point in the simulation, and we must keep in mind that some civilizations may convert multiple times, if they happen to be between members of both Group 1 and Group 2, and their social ties to a group take a few turns to consolidate. The growth

Table 3.6 Denominational census of 44,100 civilizations

Turn	Group 1	Group 2	Group 3	Group 4	Group 5	Group 6	Conversions
0	7,451	7,215	7,316	7,434	7,425	7,259	0
10	11,776	8,199	5,976	6,114	6,043	5,992	19,708
20	13,920	8,460	5,355	5,563	5,432	5,370	25,524
30	15,204	8,626	5,019	5,189	5,079	4,983	28,725
40	16,033	8,611	4,839	4,971	4,861	4,784	30,767
50	16,677	8,588	4,669	4,852	4,676	4,638	32,205
60	17,210	8,562	4,491	4,743	4,567	4,527	33,276
70	17,559	8,571	4,423	4,654	4,468	4,425	34,021
80	17,966	8,573	4,302	4,549	4,374	4,336	34,723
90	18,321	8,547	4,216	4,484	4,286	4,246	35,308
100	18,622	8,545	4,132	4,425	4,217	4,159	35,825

rate for Group 1 slows after the beginning, but does not halt. In contrast, Group 2s growth halts around turn 30, and it begins to lose out to Group 1, but remains larger than it began even after 100 turns.

If we examined the data in fine detail, we would find a few cases in which a member of a group that is active recruiting actually converts to one of the other groups. If that simulated civilization does not have any fellow members in the immediate neighborhood, developing social bonds with two or more members of some other group could lead to defection. This simulation does not automatically cluster members of an aggressive recruiting religious denomination, as the 1987 simulation did, although the software offered that option. At the opposite extreme, a simulated civilization that has no fellow members among the immediate neighbors, may wind up being socially isolated if none of the neighbors belong to either of the recruiting denominations. Table 3.7 reports the changing structure of social relations over the first 100 turns, which we could imagine might be 100,000 years for the typical inhabited planet.

A minor point to keep in mind is that the simulated civilizations on the edge of the galaxy have fewer neighbors than those inside. But 134,748 is almost exactly 3 for each of the 44,100 inhabitants, and since each bond connects two people, that implies an average effective number of bonds per person of 6, which is a density of 75%. Social isolates are the civilizations that are out of communication with any of the others, and we see that the extent of isolation jumps from 185 to 1,043 very quickly, before dropping, initially quickly then gradually. What happened was that civilizations surrounded by entirely alien religious cultures broke off communication, but then were progressively coaxed back into contact by one of the two evangelical religions.

Figure 3.2 shows the exact center of the galaxy, the same 400 civilizations as in Fig. 3.1, but now with a very different pattern of religious categorization. Group 1 is represented by blue squares, and Group 2 by light green squares. The overwhelming

Table 3.7 Social bond census of 44,100 civilizations

Turns	0-Bond Triads	1-Bond Triads	2-Bond Triads	3-Bond Triads	Total Bonds	Social Isolates
0	21,900	65,494	65,496	21,834	87,660	185
10	24,052	70,039	22,282	58,351	95,957	1,043
20	14,804	62,519	17,840	79,561	111,447	472
30	10,991	56,439	16,255	91,039	119,761	279
40	9,211	52,439	15,493	97,581	124,449	233
50	8,288	50,212	14,242	101,982	127,332	194
60	7,453	47,746	13,947	105,578	129,916	168
70	7,041	46,613	13,533	107,537	131,209	159
80	6,638	45,116	13,300	109,670	132,736	144
90	6,370	43,935	12,936	111,483	133,943	136
100	6,211	42,992	12,837	112,684	134,748	133

majority of civilizations are now connected to other nearby civilizations in the same religion category, but even just in this small part of the galaxy we see several clusters of like-minded civilizations that are largely separate from the social network that could unite the 44,100 under one faith.

Three of the 400 civilizations represented in Fig. 3.2 have no connections to any neighbors, which offers another possible resolution of the Fermi paradox. As it happens, all three had connections in Fig. 3.1, but quickly abandoned them, having different types of religion from the other civilizations to which they were connected. This might remind us of the unconventional view that extraterrestrials have visited Earth in the past, but for some reason abandoned us (von Däniken 1971, 1972; Wilson 1970; Bainbridge 1981). Perhaps we rejected them! However, this particular simulation assumes that the civilizations are at the same high level of technological development and decide whether to communicate by radio with their neighbors on largely cultural grounds.

3.9 Conclusion

This chapter and the previous one updated some admittedly early computer simulation research by the author, to provide an academic basis for consideration of a variety of virtual planets and galaxies in later chapters, beginning in Chap. 4 with some connected to academia through their educational application, and then exploring more widely across commercial simulations marketed as games. Simulations aimed at the general public tend to adopt the model of space exploration followed in the real universe by NASA, and in the cultural universe though science fiction. That is to say that they assume physical travel by human beings over cosmic distances, rather than disembodied communication via advanced information technologies. The group-

Fig. 3.2 The center of a galaxy, after cultural consolidations

modeling strategy simulations described in Chap. 5 have many similarities with the principles followed in this chapter and the previous one, while the three concluding chapters introduce a variety of role-playing simulations in which the focus is on an individual who personally experiences other planets and voyages swiftly across interstellar distances. Rather than describe one approach as realistic, and another as fanciful, we can operate under the assumption that each offers distinctive insights into particular dimensions of the cosmic challenges that humanity faces.

It is easy to imagine a variety of different ways to model cultural drift, that could have implications for the human future in our galaxy. Methods like those presented in the previous chapter could simulate a society's opinion leaders communicating about the value of nuclear energy technologies, for example constantly taking account of the economic and social benefits of innovations, both large and small, to decide which projects to invest in next. The relevance to spaceflight is that only some outcomes would prepare a good basis for development of nuclear-powered launch vehicles, given that fission-powered rockets are environmentally dangerous, and fusion-powered rockets might be less dangerous but require extensive development of technologies that do not currently exist. A very different challenge would be developing the theoretical basis for simulation of the effect on the cultures of our world if radio communications from intelligent extraterrestrials were actually detected, let alone deciphered.

Conversely, if space is the *final frontier*, and space-related progress stalls, we could be facing a time when all forms of scientific and technological innovation grind to a halt. The classic Frontier Theory was developed over a century ago by Frederick Jackson Turner, who wrote at the time when the frontier of the Wild West was closing, attributing modern democracy and free economies to the stimulating effect of a chaotic and valuable frontier, implying that society could revert to tyranny and evil exploitation without a frontier (Turner 1920). Decades later, Vannevar Bush called science the *endless frontier* (Bush 1945). But in 1996, John Horgan argued that we have reached the *end of science* (Horgan 1996). The simulation of Jet Propulsion Laboratory described in the previous chapter was very simplistic, useful primarily for educational purposes early in the development of computational expertise. Far more complex and sophisticated would be simulations of the entire scientific-technological complex of multiple disciplines and institutions, that could help us explore the implication of continued versus halted progress.

References

Abbott, Edwin A. 1884. *Flatland: A Romance in Many Dimensions*. London: Seeley.

Bainbridge, William Sims. 1976. *The Spaceflight Revolution: A Sociological Study*. New York: Wiley-Interscience.

Bainbridge, William Sims. 1981. Chariots of the Gullible. In *Paranormal Borderlands of Science*, ed. Kendrick Frazier, 332–347. Buffalo: Prometheus.

Bainbridge, William Sims. 1982. Religions for a Galactic Civilization. In *Science Fiction and Space Futures*, ed. Eugene M. Emme, 190–191. San Diego: American Astronautical Society.

Bainbridge, William Sims. 1983. Attitudes Toward Interstellar Communication: An Empirical Study. *Journal of the British Interplanetary Society* 36: 298–304.

Bainbridge, William Sims. 1984. Computer Simulation of Cultural Drift: Limitations on Interstellar Colonization. *Journal of the British Interplanetary Society* 37: 420–429.

Bainbridge, William Sims. 1986. *Dimensions of Science Fiction*, 139. Massachusetts: Cambridge Harvard University Press.

Bainbridge, William Sims. 1991. *Goals in Space: American Values and the Future of Technology*. Albany, New York: State University of New York Press.

Bainbridge, William Sims. 1995. Neural Network Models of Religious Belief. *Sociological Perspectives* 1995 (38): 483–495.

Bainbridge, William Sims. 2006. *God from the Machine: Artificial Intelligence Models of Religious Cognition*. California, Walnut Grove: AltaMira.

Bainbridge, William Sims. 2011. Direct Contact with Extraterrestrials via Computer Emulation. In *Civilizations Beyond Earth: Extraterrestrial Life and Society*, ed. Douglas A. Vakoch, and Albert A. Harrison, 191–202. New York: Berghahn.

Bainbridge, William Sims. 2015. *The Meaning and Value of Spaceflight*. Berlin: Springer.

Bainbridge, William Sims. 2017. *Dynamic Secularization*. London: Springer.

Bainbridge, William Sims. 1985. Cultural Genetics. In Religious Movements, ed. Rodney Stark. New York: Paragon.

Bainbridge, William Sims. 1997. The Omicron Point: Sociological Application of the Anthropic Theory. In Chaos and Complexity in Sociology: Myths, Models and Theory ed. by Raymond A. Eve, Sara Horsfall, and Mary E. Lee, pp. 91--101 Thousand Oaks, California: Sage Publications.

Barrow, J.D., and F.J. Tipler. 1988. *The Anthropic Cosmological Principle*. Oxford: Oxford University Press.

Blau, Peter M. 1964. *Exchange and Power in Social Life*. New York: Wiley.

Bond, Alan, and Anthony R. Martin. 1978. A Conservative Estimate of the Number of Habitable Planets in the Galaxy. *Journal of the British Interplanetary Society* 1978 (31): 411–415.

Bond, Alan, R. Anthony, and A. Martin. 1980. Conservative Estimate of the Number of Habitable Planets in the Galaxy — Part 2: Defense and Revision of the Estimate. *Journal of the British Interplanetary Society* 33: 101–106.

Bush, Vannevar. 1945. Science, the Endless Frontier. Washington DC: U.S. Government Printing Office.

Cameron, A.G.W. (ed.). 1963. Interstellar Communication New York: Benjamin.

Cavalli-Sforza, L.L., and M.W. Feldman. 1981. *Cultural Transmission and Evolution*. Princeton, New Jersey: Princeton University Press.

Clarke, J.N. 1981. Extraterrestrial Intelligence and Galactic Nuclear Activity. *Icarus* 46: 94–96.

Clarke, Arthur C. 1999. *Greetings, Carbon-Based Bipeds!: Collected Essays, 1934—1998*, 29. New York: St. Martin's.

Clarke, Arthur C. 1964. Profiles of the Future. New York: Bantam.

Cox, L.J. 1976. An Explanation for the Absence of Extraterrestrials on Earth. *Quarterly Journal of the Royal Astronomical Society* 17: 201–208.

de Magalhães, João Pedro. 2016 A Direct Communication Proposal to Test the Zoo Hypothesis. *Space Policy* 38: 22–26.

Dick, Steven J. 2003. Cultural Evolution, the Postbiological Universe and SETI. *International Journal of Astrobiology* 2 (1): 65–74.

Hahn, Matthew W. 2003. and R. Alexander Bentley, Drift as a Mechanism for Cultural Change: An Example from Baby Names. *Proceedings of the Royal Society B* 270: S120–S123.

Hanson, Robin. 1998. Burning the Cosmic Commons: Evolutionary Strategies for Interstellar Colonization. mason.gmu.edu/~ rhanson/filluniv.pdf.

Hart, M.H. 1975. An Explanation for the Absence of Extraterrestrials on Earth. *Quarterly Journal of the Royal Astronomical Society* 16: 128–135.

Heinlein, Robert A. 1941. Universe. *Astounding Science-Fiction* 27 (3): 9–42.

Hogan, James P. 1982. *Voyage from Yesteryear*. New York: Ballantine.

Homans, George C. 1950. *The Human Group*. New York: Harcourt, Brace and World.

Homans, George C. 1967. *The Nature of Social Science*. New York: Harcourt, Brace and World.

Homans, George C. 1974. *Social Behavior: Its Elementary Forms*, 16. New York: Harcourt, Brace, Jovanovich.

Homans, George C. 1984. *Coming to My Senses: The Autobiography of a Sociologist*. New Brunswick, New Jersey: Transaction.

Horgan, John. 1996. *The End of Science: Facing the Limits of Knowledge in the Twilight of the Scientific Age*. Reading, Massachusetts: Addison-Wesley.

Jones, E.M. 1976. Colonization of the Galaxy. *Icarus* 28: 421–422.

Jones, E.M. 1978. Interstellar Colonization. *Journal of the British Interplanetary Society* 31: 103–107.

Jones, E.M. 1982. Estimates of Expansion Time Scales. In *Extraterrestrials: Where Are They?*, ed. M.H. Hart, and B. Zuckerman, 66–76. New York: Pergamon.

Keyfitz, Nathan. 1982. Choice of Function for Mortality Analysis: Effective Forecasting Depends on a Minimum Parameter Representation. *Theoretical Population Biology* 21: 329–352.

Lewis, C.S. 1996. Perelandra, p. 70. New York: Scribner's.

Lissauer, Jack J. 2000. How Common are Habitable Planets? NASA Technical Reports Server.

Lumsden, C.J., and E.O. Wilson. 1981. *Genes, Mind, and Culture Cambridge*. Massachusetts: Harvard University.

MacArthur, Robert H., and Edward O. Wilson. 1967. *The Theory of Island Biogeography Princeton*. New Jersey: Princeton University Press.

MacGowan, Roger A., and I. Frederick. 1966a. *Ordway, III, Intelligence in the Universe*. Englewood Cliffs, New Jersey: Prentice-Hall.

MacGowan, Roger A., and I. Frederick. 1966b. *Ordway, Intelligence in the Universe*. Englewood Cliffs, New Jersey: Prentice-Hall.

Miller, Neil E., and John Dollard. 1941. *Social Learning and Imitation*. New Haven: Yale University Press.

Molton, P.M. 1978. On the Likelihood of a Human Interstellar Civilization. *Journal of the British Interplanetary Society* 31: 203–208.

Newman, W.I., and C. Sagan. 1981. Galactic Civilizations: Population Dynamics and Interstellar Diffusion. *Icarus* 46 (293–327): 293.

Schwartzman, D.W. 1977. The Absence of Extraterrestrials on Earth and the Prospects for CETI. *Icarus* 32: 473–475.

Singer, C.E. 1982. Galactic Extraterrestrial Intelligence. *Journal of the British Interplanetary Society* 35: 99–115.

Skinner, B.F. 1938. *The Behavior of Organisms*, 21. New York: Appleton-Century.

Sklovskii, I.S., and Carl Sagan. 1966. *Intelligent Life in the Universe*. New York: Dell.

Smith, A.G. 1981. Constraints Limiting the Rate of Human Expansion into the Galaxy. *Journal of the British Interplanetary Society* 34: 363–366.

Stull, M.A. 1979. On the Significance of the Apparent Absence of Extraterrestrials on Earth. *Journal of the British Interplanetary Society* 32: 221–222.

Swift, David W. 1990. *SETI Pioneers: Scientists Talk about their Search for Extraterrestrial Intelligence*. Tucson: University of Arizona Press.

Tang, T.B. 1982. Fermi Paradox and C.E.T.I. *Journal of the British Interplanetary Society* 35: 236–240.

Thomas, Shirley. 1968. Men of Space, vol. 8, pp. 14—20. New York: Chilton.

Thomas Howard, C.S. 1987. Lewis Man of Letters. Worthing, England: Churchman Publishing. unpaginated.

Tinsley, B.A. 1980. Technical Development and Colonization as Factors in the Long-Term Variation in Limits to Growth. *Cosmic Search* 2 (4): 10–12.

Tipler, F.J. 1980. Extraterrestrial Intelligent Beings do not Exist. *Quarterly Journal of the Royal Astronomical Society* 21: 267–281.

Tipler, F.J. 1981a. A Brief History of the Extraterrestrial Intelligence Concept. *Quarterly Journal of the Royal Astronomical Society* 22: 133–145.

Tipler, F.J. 1981b. Additional Remarks on Extraterrestrial Intelligence. *Quarterly Journal of the Royal Astronomical Society* 22: 279–292.

Tipler, F.J. 1982. Extraterrestrial Intelligence: The Debate Continues. *Physics Today* 18: 26–38.

Turner, Frederick Jackson. 1920. The Frontier in American History. New York: Holt.

van Vogt, A.E. 1942. Asylum. *Astounding Science-Fiction* 29 (3): 8.

Viewing, D. 1975. Directly Interacting Extra-Terrestrial Technological Communities. *Journal of the British Interplanetary Society* 28: 735–744.

von Däniken, Erich. 1971. *Chariots of the Gods?*. New York: Bantam Books.

von Däniken, Erich. 1972. *Gods from Outer Space*. New York: Bantam Books.

von Hoerner, Sebastian. 1975. Population Explosion and Interstellar Expansion. *Journal of the British Interplanetary Society* 28: 691–712.

Walters, Clifford, Raymond A. Hoover, and R.K. Kotra. 1980. Interstellar Colonization: A New Parameter for the Drake Equation? *Icarus* 41: 193–197.

Wilson, Edward O. 1975. *Sociobiology, the New Synthesis*, 551. Cambridge, Massachusetts: Harvard University Press.

Wilson, Clifford .1970. *Crash Go the Chariots*. New York: Lancer.

Chapter 4
Educational Simulations of the Evolution of Spaceflight

Currently, the inhabitants of planet Earth cannot directly experience either the environment of another planet or the social movement that might actually take humans there in some future century. Some astronomy simulations, like *Starry Night*, have become popular with amateur astronomers and have some use in schools (Bainbridge 2015). Thus we may hope that realistic spaceflight simulations might become widely used in academic curricula, at the college level as well as in high school and earlier grades. We can imagine them incorporating social science principles and extending the actual history of astronautics into plausible representations of future development stages. Where now has educational spaceflight software reached, if humans are indeed destined to navigate their difficult path to the stars?

4.1 Realistic Educational Simulations

A very wide variety of educational software exists, that teaches fundamental facts of astronomy and principles of spaceflight, but social dimensions are usually lacking. For example, NASA offers a well-made but limited simulation of the Curiosity Mars rover: "Welcome to Experience Curiosity, a WebGL tool to learn about the and its adventures in the Pahrump Hills region of Gale Crater on Mars. Explore the highlights of the Pahrump Hills area, replay some of Curiosity's activities, or take control and use a virtual rover to have a look around."[1] The reference to WebGL means that the software smoothly runs inside any common web browser, and does not require any special download action by the user.[2] Figure 4.1 shows this virtual Curiosity in the midst of one of seven simulations of real activities the rover had performed.

[1] https://eyes.nasa.gov/curiosity, accessed June 2017.

[2] https://en.wikipedia.org/wiki/WebGL, accessed June 2017.

W. S. Bainbridge, *Computer Simulations of Space Societies*, Space and Society,
https://doi.org/10.1007/978-3-319-90560-0_4

Fig. 4.1 Operating a virtual Curiosity Rover on Mars

The virtual environment is subjectively rather small, but its scope is expanded by the incorporation of many real images the rover had sent back to Earth. The control menu offers four main choices, two of which involve learning about the computer-generated rover itself. In Learn About Curiosity, the user can select one of its 10 main components, see it highlighted in the model, and read a paragraph describing its function. In Control Curiosity, the user can move three of those components: the robotic arm, one of the antennas, and the mast holding two cameras, a telescope, and a spectroscope. The two other main choices drive the rover around and reprise the seven activities:

Free Drive and Rover Replay:
Sol 826 Panorama
Sol 908 Drill
Sol 868 Selfie
Pahrump Hills Highlights:
Sol 751 Approach
Sol 771 Pink Cliffs
Sol 796 Whale Rock
Sol 929 Garden City

The activity shown in Fig. 4.1 is the Sol 868 Selfie, *sol 868* meaning the 868th Martian day after landing, and *selfie* meaning that the rover is taking pictures of itself. The picture at the lower right of the image shows one part of the selfie, which is an assembly of images called a *mosaic*. The interface window at the left of the image offers the choices and currently displays part of the following text:

This self-portrait shows Curiosity at the "Mojave" site, where its drill collected the mission's second taste of Mount Sharp. The scene combines dozens of images taken during January

2015 by the Mars Hand Lens Imager (MAHLI) camera at the end of the rover's robotic arm. The view does not include the rover's robotic arm. Wrist motions and turret rotations on the arm allowed MAHLI to acquire the mosaic's component images. The arm was positioned out of the shot in the images, or portions of images, that were used in the mosaic.

The user can view Curiosity from any angle or distance, then click *replay* to make it go through the sequence of moves that was actually used to take the images that combined to form the selfie. This very well illustrates the flexibility of the arm, and helps a student understand how the rover operated. In any of the seven activities, a message will display briefly explaining that events will take place much more rapidly on the computer screen than they really did on Mars, in this case: "Playback approximately 60X actual speed." Experience Curiosity does not, therefore, simulate how the real Curiosity was controlled.

When the Soviet space program sent two Lunokhod rovers to the Moon in 1970 and 1973, they were in fact controlled in realtime by teams of five technicians.[3] This was possible because the Moon is close enough that signals can complete the round trip between rover and controllers in about three seconds. Even at its closest approach to earth, the round trip at the speed of light to and from Mars takes more than six minutes, and the typical time would be much longer. The only circumstance in which we can imagine realtime operation of Mars robots by humans would be in the context of a colony that performed much of its construction and resource gathering work through teleoperated machines, rather than manually by people wearing spacesuits. The humans could sit comfortably inside their pressurized habitats, and the machines would operate in the nearby desert.

Sociologist Janet Vertesi of Princeton, author of *Seeing Like a Rover: How Robots, Teams, and Images Craft Knowledge of Mars*, has studied intensively through direct observation how teams at Jet Propulsion Laboratory interpreted the images returning from the first generation of Mars rovers, not merely in the performance of scientific analysis but also in planning the future moves and data collection activities of the rovers (Vertesi 2014). She is currently preparing publications on her studies of the teams operating the Cassini orbiter of Saturn, and in the early stages of research at JPL on similar aspects of the future Europa mission. The training of the high-tech real-world rover and orbiter teams is comparable to social simulations, and employs extensive computation. Indeed, team members are constantly learning, developing their comprehension of the Martian environment collectively, communicating through and around a toolkit of computer technologies that offer multiple perspectives.

On her academic website, Vertesi says, "My work is mostly ethnographic, although I am also trained in ethnomethodology, and I especially enjoy applying my sociological insights to technology development through the field of Human-Computer Interaction."[4] Ethnomethodology is a sociological method and school of thought that emerged in the 1960s, based on phenomenology in philosophy, and seeking to understand how people conceptualize their world and their location in it

[3]https://en.wikipedia.org/wiki/Lunokhod_programme, accessed December 2017.

[4]https://sociology.princeton.edu/faculty/janet-vertesi, accessed December 2017.

Garfinkel (1967); Schutz (1967). Thus ethnomethodology is comparable to cognitive science, but as is the case for sociology more generally, has not been effectively integrated with that more recent convergence of fields concerning human perception and cognition.

Vertesi explains, "Scientific seeing is not a question of learning to see without bias, Instead, scholars of scientific observation remind us, it entails acquiring a particular visual skill that allows a scientist to see some features as relevant for analysis and others as unimportant (Vertesi 2014)." This raises a crucial question about educational computer simulations. Is the computer merely a tool for delivering a pre-determined curriculum, like training in the multiplication table or periodic table of the chemical elements, essentially using games to motivate students to memorize the facts their teacher thinks they need to know? Or should educational computer games better be conceptualized as laboratories where students may experiment, following their own proclivities. Perhaps it is better to conceptualize space-related computer games as wilderness territories where students may become explorers. On a virtual Mars they will not learn astronomy but gain personally-relevant skills at improvising inside information systems or environments from which abilities may transfer back to their real terrestrial lives.

Experience Curiosity is designed for use by casual visitors rather than school classes. NASA has a very different site, SpacePlace, offering games and other activities suitable for classroom use, plus information and tools to help teachers integrate the offerings into their curriculum.[5] Each item seems rather simple, and unlikely to be the basis of serious research on educational information technologies, so I decided to focus instead on three educational game research projects sponsored by the National Science Foundation, *Lunar Quest, Race to Mars*, and *Quest Atlantis*. I had already explored *Quest Atlantis* for a few hours, years before, and the point now was to gain some historical perspective. However the result was unexpected, and, frankly, distressing. The creators of these three educational game systems were both visionary and competent, and yet the educational success of the projects was less than might have been hoped. However, conceptualized as research projects, they may have contributed significantly to our understanding of the educational dimensions of computer simulations.

Partially funded by grants from the National Science Foundation, *Lunar Quest* was intended "to teach introductory physics concepts to college students" or was "designed to support learning of physics concepts."[6] However, those grants were made back in 2006 and 2009, and after collecting the data for many publications based on the research, the team of scientists and programmers had to move on to other projects, and the website at the University of Central Florida that provided *Lunar Quest* no longer exists. Searching for "lunar quest" online will mostly turn up false hits, because several other things have similar names, but a video was posted October 13, 2010, by Tim Holt, a member of the team, who noted: "The game has

---

[5]https://spaceplace.nasa.gov/menu/parents-and-educators, accessed December 2017.

[6]www.nsf.gov/awardsearch/showAward?AWD_ID=0537078; www.nsf.gov/awardsearch/showAward?AWD_ID=0856045, accessed December 2017.

a retro-future style, and was a project at UCF to do game-based learning of physics concepts." It shows a man inside a pressurized lunar base, then walking around the surface in a space suit, viewed only 101 times by July 9, 2017. It offers a link to a site that had more information, but has since vanished.

One of the findings of the *Lunar Quest* research shows how science can be successful when it discovers problems, as well as when it discovers solutions. An implicit issue throughout this book is the relationship between the physical science of the universe and possible future human social activities beyond the boundaries of Earth. The *Lunar Quest* team encountered interesting problems in the attempt to render physics education social through a Massively Multiplayer Online Game (MMOG):

> It was quickly discovered, however, that the MMOG environment had limitations that made teaching physics more difficult than the single player environment. It is difficult to model realistic physics in a MMOG due to network latency issues; it is difficult to ensure an identical learning experience between players due to the ability for players to assist each other; and it is difficult to grade a player on a deeper level than pass/fail due to the stringent questing rules found in an MMOG. This is why the design team incorporated mini-games into Lunar Quest to deliver the learning content. Smith and Sanchez (2010)

These mini-games would be experienced solo, thus separating the physics education from the social interaction. *Race to Mars* placed solo education in the context of wider social issues. NASA promoted this project, but NSF provided some of the funding in a Small Business Innovation Research (SBIR) grant in 2011. Its goal was to develop "a problem-based serious game to interest and engage high school females in aerospace engineering and STEM challenges related to flight and space. Deliverables include a framework of game design principles to attract underrepresented groups, a serious game module, and teacher and student guides. Our hypothesis is that by identifying new ways to design, develop, and implement engineering curriculum using serious games with game elements that engage underrepresented groups, we can increase interest, motivation, and learning outcomes in engineering for high school students (particularly for females and minority females)."[7] SBIR grants are intended to transfer technology from academia to industry, through assisting start-up companies, in this case WisdomTools Enterprises in Indiana.

The WisdomTools website makes *Race to Mars* sound rather social: "As a pioneer of space exploration, you are being asked to start your own shipping company to transport materials from the Moon to Mars. To get to Mars, you'll need to sign a contract, find crewmembers, and build a spacecraft. Remember to pick your team and design your spacecraft to fit the specific goals of your contract!"[8] But the download tool on the website does not work, and neither does the link to the support stage. The home page suggests that some of the company's products can be bought through the widely used online store, Steam. However, the game called *Race to Mars* on Steam seems to be from a Polish developer, and the game connected to WisdomTools seems to be *Starlite*, described by Wikipedia as "a multiplayer online game which, since November 2009, is being developed by Project Whitecard Studios and WisdomTools

[7] www.nsf.gov/awardsearch/showAward?AWD_ID=1047122, accessed December 2017.

[8] www.wisdomtools.com/aeroengineer-race-to-mars, accessed June 2017.

Fig. 4.2 A Mars rover being operated from a large surface transport

Enterprises (now defunct). The game world will be set in the near future with the ability to explore planned and possible near-future planetary missions, which is facilitated by the use of NASA Learning Technologies, and Innovative Partnerships programs."[9]

The game available on Steam was *Starlite: Astronaut Rescue*, "developed in collaboration with NASA," which launched January 27, 2014 and cost me only $2.99 on July 9, 2017. Its advertisement in the Steam store described it as only the first step toward a major social experience in space: "Players navigate a future Mars mission in which they must construct a habitat, craft tools and use advanced robots. The game contains hands-on science inquiry and problem solving in mathematics, physics and engineering. Starlite: Astronaut Rescue is the first release in the series leading up to alpha testing of the multiplayer online game Starlite: Astronaut Academy in summer 2014." However, I find no evidence online that *Starlite: Astronaut Academy* ever launched. For example, it is not listed among the fully 1,263 multi-player games on the MMORPG.com blogsite.[10] Figure 4.2 shows a late stage in this solo-player game in which the user navigates a rover named Sam 2 to a target location.

Despite the extra time required to take 108 screenshots to document everything, and a confusion at one point about how the user interface worked, *Starlite: Astronaut Rescue* took me just 31 min to complete. The user's avatar is standing inside

[9]https://en.wikipedia.org/wiki/Starlite_(video_game), accessed June 2017.

[10]www.mmorpg.com/games-list, accessed June 2017.

the huge vehicle in the background, initially getting instructions from a non-player astronaut, then undertaking a mission to locate the crashed ship of another astronaut. To triangulate the source of the emergency signal, two rovers must be driven to the designated targets, then turn until the signal turns green in color. Prior to that, a rather arcane calculation must be done by the student to decide the length of the antennas the rovers should carry, in order to focus narrowly on the signal at its estimated wavelength. The Steam site indicated that perhaps as few as 176 people had purchased this brief game, of whom 139 rated it, only 41% of them positively.

These two games indicate how difficult it may be to create social simulations that both educate the users in physics or astronomy, and are attractive to the wide, young audience for computer games. A rather more successful example approached the problem from a rather different direction. Sasha Barab at Indiana University got an NSF Small Grant for Exploratory Research grant in 2002, $87,316 to support the first stage of the Quest Atlantis Project "to create exciting opportunities for children in an after school context."[11] Then, on this basis of this pilot study, in 2004 Barab received a $1,704,440 grant, Quest Atlantis: Advancing a Socially-Responsive, Meta-Game for Learning: "This grant will undertake a series of naturalistic studies and experimental manipulations that will advance theoretical and practical knowledge with respect to the design of multi-user virtual environments for supporting science learning."[12] He and his research team developed a theory of *transformational play*, identifying three key factors that could render computer games truly educational:

> Person With Intentionality (positioning players as protagonists with the responsibility of making choices that advance the unfolding story line in the game)
>
> Content With Legitimacy (positioning the understanding and application of academic concepts as necessary if players are to resolve the game-world dilemmas successfully)
>
> Context With Consequentiality (positioning contexts as modifiable through player choices, thus illuminating the consequences and providing meaning to players' decisions)[13]

There were several social aspects of *Quest Atlantis*, as demonstrated on its YouTube channel, notably the fact that it was designed for collective use by school classes www.youtube.com/channel/UCmCLe8lYlrC34gZc-SWqblw. Much of the extensive educational content concerns the global environment, which means both preservation of nature and international fellowship. It is really a cluster of virtual worlds, rather than just one:

> By moving their avatar through the on-screen environment, citizens travel to virtual worlds where they can read about and listen to the themes of these worlds, complete quests, talk with other children and with mentors, and build their virtual personae. The virtual space - OTAK - is divided into worlds. Each of the four primary worlds of the OTAK - (a) Unity World, (b) Culture World, (c) Ecology World, and (d) Healthy World - is divided into three villages (e.g., Unity World includes Global Village, Community Power Village, and All About Us Village) that hold up to 25 quests. The OTAK currently contains the OTAK-Hub, and the four primary worlds. The OTAK-Hub is the central location from which to teleport to each

[11] www.nsf.gov/awardsearch/showAward?AWD_ID=0137298, accessed December 2017.

[12] www.nsf.gov/awardsearch/showAward?AWD_ID=0411846, accessed December 2017.

[13] Transformational Play: Using Games to Position Person, Content, and Context (Barab et al. 2010).

> of the worlds. It also offers introductory quests, the Trading Post, general information, and the Quests of the Moth. Each village has a title reflecting a theme - for example, Community Power, Animal Habitats, Water Quality, Words of Meaning - and an associated series of engaging quests. The themes were designed to span diverse areas of knowledge and feature something for almost everyone, yet still overlap academic categories. Each village houses a spectrum of quests (engaging academic tasks that take 20 min - 1 week to complete) ranging from simulation to application problems of varying levels of complexity (Barab et al. 2005).

In many respects, *Quest Atlantis* was quite excellent. However, in the context of this book published years after *Quest Atlantis* launched, two caveats must be mentioned. First, neither it nor more recent educational social computer games have become major elements of standard educational practice. Whether that is evidence about their limitations, or about the obsolescence of traditional schools, cannot be decided here. Second, the virtual worlds are not described as planets. They could be, because while resembling Earth they are not depicted as being on our world, and the ambiguity of their cosmic location may reflect the fact that all worlds which humans are happy to imagine are metaphors for the planet we inhabit.

4.2 Space Program Strategies

A number of so-called "strategy games" depict space-related environments where the player makes a series of choices to build up a system of resources and capabilities to the point that a difficult space-related goal is achieved. A famous example that has educational intentions is *Buzz Aldrin's Race Into Space*, which is a simulation of the Space Race competition between the US and USSR, in which Buzz Aldrin himself played an important role as an Apollo astronaut. Launched in 1993 and simulating the space race between the US and USSR in the period 1957–1977, it can be played in three ways: (1) a player managing the American space program competing with the computer playing the Soviet space program, (2) a player taking the Soviet side, and the computer, the American, or (3) two players playing the competing roles. The player is the manager of a space program, taking turns in making investment decisions and launching a series of ever more ambitious space missions. The winner is the side that reaches the Moon first. Aside from brief animations that depict events at special points in the game's progress, it is visually simple.

While a commercial game and not explicitly educational, it does seek to educate players both about the actual history of spaceflight and the strategic logistics required to build a complex technological program. As the instruction manual says: "This is a strategy-oriented game that requires short and long-term planning. You'll need to determine what space hardware is needed to complete your objectives. While it is certainly not required, it is suggested that you read some of the historical material on the space race. The American and Soviet strategies are quite insightful (Bronner 1992, p. 1)."

A team of researchers at the University of Nevada, led by Sushil Louis, has conduced several studies in which various artificial intelligence methods were used to play strategy games (Avery and Louis 2010; Liu et al. 2013). This methodology could be described as computer simulation of people interacting with simulations, and it can be used to evaluate both the strategy game and the particular form of artificial intelligence, whether genetic algorithms or machine learning. However, these studies also illustrated how humans must learn new skills playing strategy games, that may transfer to the real world, through both improved game design and the education of players:

> Finding effective and robust strategies in Real-Time Strategy (RTS) games presents a challenging problem. RTS game players must compete for resources, build up an economy that is able to support their military force, expand their control over the map, and eventually destroy their opponent's base. Any advances in developing RTS game players will impact planning and execution in competitive industrial settings through the development of smart, realistic opponents (Ballinger and Louis 2013).

Destroying the enemy's base is neither possible nor rewarded in *Buzz Aldrin's Race Into Space*, and the map is the series of steps in technological development required to reach the Moon. Some of those steps mirror with some degree of accuracy the real historical developments, for example the US and USSR development of comparable manned vehicles: Mercury and Vostok, Gemini and Voskhod, and Apollo and Soyuz. However, the latter pair of vehicles could be supplemented or supplanted by a pair of mini shuttles, thus taking space history on a somewhat different course. The instruction manual explains that there are twenty different ways to perform the actual lunar mission, and urges players to have contingency plans in case the chosen set of technological steps turns out to be too difficult. Throughout, each mission is associated with a particular degree of risk, and the safest route may also be a slow one, implying that winning the game is usually associated with taking moderate risks. Thus the game teaches theories of human action, as well as the history of real spaceflight and the alternative choices that were not actually taken.

In October 2014, the Steam online game store released a sequel, *Buzz Aldrin's Space Program Manager*, that Aldrin himself had helped develop, so I explored it, beginning June 30, 2017. By near coincidence, that was the day US president Donald Trump announced the revival of the National Space Council, which had existed from 1989 until 1993, and Aldrin was present for the ceremony, standing next to Trump and interacting with him. As Trump was signing his executive order, Aldrin commented, "Infinity and beyond."[14] This is the paradoxical motto of Buzz Lightyear, a cartoon character with whom Aldrin has identified.[15] It is worth noting that the central character in the 1950–1955 television series, *Space Patrol*, was also named Buzz, and played by an actor who had been a fighter pilot in World War II.[16] The entanglement of fact and fiction in the history of spaceflight is one of its defining features, an expression of visionaries who do wish to voyage beyond infinity.

---

[14]www.whitehouse.gov/the-press-office/2017/06/30/remarks-president-signing-executive-order-national-space-council, accessed December 2017.

[15]https://en.wikipedia.org/wiki/Buzz_Lightyear, accessed June 2017.

[16]https://en.wikipedia.org/wiki/Space_Patrol_(1950_TV_series); https://en.wikipedia.org/wiki/Ed_Kemmer.

The most popular space games available on Steam connect to popular movie franchises, for example *Alien: Isolation* launched in October 2014 that got 13,169 reviews by players, 91% of them positive. *Buzz Aldrin's Space Program Manager* is by no means an especially popular game, receiving 282 player reviews on Steam, 227 or 80% of them positive. Among the most recent posts, Ph4nt0 m said, "Great game! Not for everyone but really reminds me of Buzz Aldrin's Race Into Space that I had on floppy disks when I was a kid." One of the game's developers replied, "Thanks for the review! Glad you are enjoying the game and that it's bringing you fond memories from the original BARIS :)." HistoryNet published a serious review that considered it a popular summary of spaceflight history, but commenting, "Some may feel this game is less about space than shuffling staff between training and programs."[17] SpaceSector, a blogsite for space and science fiction strategy games, praised the new feature that made this game a serious simulation of space-related social behavior:

> One of the best additions to the game is the incorporation of scientists, engineers, technicians and mission controllers as actual characters with stats, wages and so on rather than just an arbitrary button to click like in BARIS. Here, you actually need to choose who to hire based on their skills, which can be further refined by advanced training, then assigning them to different programs best suited for them. Mission control is the exact same, but instead you assign the different controllers to the various stations on the launch of the craft, which then serve as modifiers to the various success rolls that occur during a mission... This might be a relatively minor feature in the greater scheme of things, but it adds on an entirely new layer of strategy onto the management of personnel; do you keep your scientists at work with their current skill levels, or do you send them to advanced training, losing access to their skills for a few seasons while they improve, giving you better research in the long run?[18]

A very substantial part of this simulation places the user exactly in the role of program manager, selecting, training, and directing computer simulations of people. Figure 4.3 shows the default view of the NASA Space Complex and the buildings of its seven main organizational units: (1) Astronauts Centre, (2) Public Affairs Office, (3) Headquarters, (4) Mission Control Centre, (5) Museum, (6) SET Centre, and (7) Vehicle Assembly Building. Using the interface associated with this image, the user can take action within each of these units.

Perhaps most interesting is SET, the office of Scientists, Engineers and Technicians, at the center and to the left of the dark box that offers a description of their function, saying they "are the ones that RD mission components in order to raise their reliability and increase the chances of success when used in a mission." Practically the first step in each run involves selecting from five SET candidates, each of whom has a specified measure of expertise in each of five fields[19]:

> Rockets: These are the vehicles that carry uncrewed payloads to outer space, such as unmanned satellites or planetary probes. Depending on their capabilities, they can be clas-

[17] Dy, Bernard. 2017. Review: Buzz Aldrin's Space Program Manager, *HistoryNet*. http://www.historynet.com/review-buzz-aldrins-space-program-manager.htm

[18] Salt, Chris. 2014. Buzz Aldrin's Space Program Manager Review. In *SpaceSector*. http://www.spacesector.com/blog/2014/11/buzz-aldrins-space-program-manager-road-to-the-moon-review

[19] Instruction manual for *Buzz Aldrin's Space Program Manager*, pp. 25–26.

Fig. 4.3 A simulated NASA space complex

sified as 'Light', 'Medium' and 'Heavy'. Mission components that belong to this group are only suitable for carrying unmanned robotic spacecraft to space. Examples of mission components that belong to this group are Juno II and the R-7 Sputnik rocket.

Space probes: This group embodies both unmanned satellites and probes sent to explore other celestial bodies. Examples of mission components that belong to this group are the Sputnik satellite and the Mars Viking lander.

Human-rated rockets: Mission components from this group are launch vehicles that have been certified as capable of carrying human beings and, as such, they tend to be more expensive than their nonhuman rated counterparts. Notice that human-rated rockets can also be used to carry space probes, a strategy that can be useful in order to raise the reliability of the rocket beyond their maximum R&D level. Examples of human-rated rockets include the Atlas booster, the Saturn V booster and the N1.

Crewed spacecraft: This group encompasses all types or spacecraft that are capable of carrying human beings on board. Examples from this group are the Vostok spacecraft and the Apollo Command and Service Module (CSM).

EVA suits: These are the pieces of equipment that allow astronauts and cosmonauts to work in space. Examples from this group include the Berkut space suit and the Lunar Roving Vehicle used in the Apollo lunar missions.

Like the original Buzz Aldrin game, this one can be a replay of the historical space race between the United States and the Soviet Union, each following the same structure but in accomplishment of different competing missions. With this in mind, I carried out a pilot study following a methodological paradigm called *algorithm audit*, systematic experiments interacting with a computer system to learn about the factors that determine its behavior. This method has become suddenly very popular in social research on computer systems, because it can often find evidence that a particular online business or other humanly consequential computer system is treating human beings unfairly (Sandvig et al. 2016). The goal here is more modest: to see if

Table 4.1 Statistics for 100 simulated scientists, engineers and technicians

	Rockets	Space probes	Human-rated rockets	Crewed spacecraft	EVA suits
Mean (%)	40.3	41.2	41.7	38.8	40.8
Standard deviation (%)	9.0	9.8	10.1	9.1	9.8
Correlations:					
Salary	0.01	0.16	0.14	−0.01	0.07
Morale	−0.42	−0.32	−0.32	−0.32	−0.37
Learning capacity	0.53	0.42	0.52	0.49	0.64
Age	0.15	0.08	0.16	0.11	0.09
Soviet	0.03	0.07	−0.06	−0.05	−0.09
SET correlations:					
Rockets	1.00	0.33	0.45	0.21	0.30
Space probes	0.33	1.00	0.45	0.16	0.30
Human-rated rockets	0.45	0.45	1.00	0.27	0.41
Crewed spacecraft	0.21	0.16	0.27	1.00	0.50
EVA suits	0.30	0.30	0.41	0.50	1.00

it is possible to find patterns in how the skills were allocated to the simulated SET experts. I started *Buzz Aldrin's Space Program Manager* 20 times, 10 times each for the US and USSR sides, recording the five skills plus other data for a total of 100 simulated people, as summarized in Table 4.1.

The SET skills were expressed as percentages, and after the start of a run some of them could be trained up to higher levels. Across the 5 skills the means and standard deviations are rather similar, averaging around 40.6 and 9.6%. The salary variable is the cost of using a particular simulated person, and it does not seem to correlate with their current skill. In contrast, morale has big negative correlations, which suggests that it was intended as a separate variable the player would consider, favoring high morale candidates because they might be more willing to work hard. Learning capacity has big positive correlations, perhaps explaining why some candidates already have high skills: They learned fast in hypothetical earlier jobs that were not part of *Buzz Aldrin's Space Program Manager*. The correlations among the five SET skills are not easy to interpret, but some are quite large. Crewed spacecraft and EVA suits must both be designed to preserve frail human bodies, so they are similar and thus correlate at 0.50. The 0.45 correlations linking human-rated rockets to both rockets and space probes suggest that humans require reliable rockets and reliable payload equipment, while the lower 0.33 correlation between rockets and probes implies they require somewhat more different expertise from each other.

4.3 Experiencing the Martian Environment

Early in the twenty-first century, Richard Childers and Bill Kovacs invested great energy and intelligence, developing a concept for a virtual Mars that would be more realistic than the existing online games, expressed through a company called Virtual Space Entertainment. After the unexpected death of Kovacs, Childers expressed their shared vision: "Perhaps in the creation of this exciting new world, we will open new vistas for education and communication. Perhaps the perspective we gain by looking at our world through the eyes of Martian colonists in the year 2150 will help us to gain a broader understanding of the Earth in the year 2009. Perhaps we can find a new Future in Mind Childers (2010)." The National Science Foundation supported "a one-year proof of concept study that includes the design, development, and research of two prototype science activities for the virtual Blue Mars Science Center located on the Blue Mars 2150 platform developed by Virtual Space Entertainment."[20] In 2012, an article in *International Journal of Game-Based Learning* reported the results: "This study provides evidence that sustained scientific inquiry can be nurtured in an MMO game and that gamers' relationships with characters in the game and other players may help facilitate that inquiry (Asbell-Clarke et al. 1989)."

However one imagines it, a virtual Mars has great educational potential, and many other visionaries have gone some distance toward developing their own simulated red planets. On May 16, 2017, the online game distribution service, Steam, offered gamers the opportunity to travel to Mars, where a substantial colony had just been mysteriously destroyed, and therefore survival would be especially difficult. In a simulation called *ROKH*, the environment was portrayed with a good deal of accuracy, and survival required much exploration and analysis, thus rendering the experience highly educational, despite the fact this was a commercial multiplayer game. *ROKH's* Facebook page proclaimed:

> It's Launch day! Today, your journey to Mars begins. The Development team is thrilled to have #ROKH out on #Steam. ROKH is a complete immersion into a cold and dry land that harbors a wealth of resources to extract and exploit. Players will explore Mars, paving the way for settlers to come, as Earth's resources are tapped and the planet is dying. Watch the new launch trailer and be sure to check out our Steam page! Thank you for agreeing to help us develop a future home for humanity on Mars. Much more to come![21]

The Steam page where the game cost $24.95 advertised: "Have you got what it takes to survive on Mars? *ROKH* is a multiplayer survival game featuring scientific realism and extremely deep crafting and building. Make the red planet your sandbox and create a thriving colony."[22] However, that Steam page warned: "This Early Access game is not complete and may or may not change further. If you are not excited to play this game in its current state, then you should wait to see if the game progresses further in development." Elsewhere, Steam explained the concept:

---

[20]www.nsf.gov/awardsearch/showAward?AWD_ID=0917520

[21]www.facebook.com/rokhthegame, accessed June 2017.

[22]https://store.steampowered.com/app/462440/ROKH, accessed June 2017.

> What is Early Access? Get immediate access to games that are being developed with the community's involvement. These are games that evolve as you play them, as you give feedback, and as the developers update and add content. We like to think of games and game development as services that grow and evolve with the involvement of customers and the community. There have been a number of prominent titles that have embraced this model of development recently and found a lot of value in the process. We like to support and encourage developers who want to ship early, involve customers, and build lasting relationships that help everyone make better games.[23]

This is to say that *ROKH* is still rather early in its development, both incomplete and bug-ridden, and does not meet the traditional standard that defined when videogames were ready to be sold. One factor influencing expectations is the state of the technology by which games are delivered to the users. In the late 1970s, I had a Fairchild F videogame system, that was social in that it had two controllers, and many of the games consisted of simple combat between crude graphic representations of vehicles, such as armored tanks in the game *Desert Fox*, fighter planes in *Spitfire*, and spaceships in *Space War*, which was actually more primitive than *Spacewar!* that was 15 years older. The technological innovations in the Fairchild F are described by Wikipedia: "It has the distinction of being the first programmable ROM cartridge-based video game console, and the first console to use a microprocessor."[24] In principle, the data and program on one of the ROM cartridges could have been downloaded from Internet, but that would not be feasible for the general public for about two decades. Thus, in the good old days games needed to be completely developed and debugged before sale, since there was no mechanism for updating them.

Today, the situation is the exact opposite. I purchased the original version of *Buzz Aldrin's Race Into Space* on a disc, but downloaded the greatly expanded recent version from Steam. Updates can be made at any time, whether modest improvements that are free or major expansions that are costly. Low-popularity games may be commercially viable on Steam, at least in terms of the distribution costs if not the expense of creating them. More generally in the gaming industry, many games are available at multiple stages in their development. *Alpha test* versions may be buggy, and a player's progress may be erased at any time, but they can be fun to explore. The meaning of *beta test* versions has become uncertain, originally designed just for free participation in the developers' test runs prior to completion, but now including soft launches and even requiring some kind of payment.

In the case of *ROKH*, the recent technological transformation of software delivery has made it possible to charge for an alpha version, but with the hope that includes the right to keep playing after a game has been fully created. Of interest for a chapter on educational simulations of space-related social behavior, there are three ways *ROKH* is educational.

First, by depicting Mars with some degree of accuracy, and offering experience setting up a virtual colony there, *ROKH* does teach the user about the real challenges facing humanity if it tries to expand beyond the Earth. One must wear a spacesuit and seek oxygen to breathe. The gravity is somewhat lower than here on Earth, but there

---

[23]https://store.steampowered.com/earlyaccessfaq, accessed June 2017.

[24]https://en.wikipedia.org/wiki/Fairchild_F_Videocarts, accessed June 2017.

are limits on the weight of resources that one may carry. The visual environment seems quite similar to the pictures received from Curiosity and the other real Mars landers. The logistics of assembling the architecture and machinery for a viable colony require great effort but also careful planning. Thus, each player is a student learning by trial and error, given that the online instruction manual and in-game hints are very incomplete.

Second, the players serve as teachers to each other, sharing their experiences and often engaging in debates about how to deal with a particular problem. One medium for their communication is a text chat system built into *ROKH*, but that is limited because one unusual feature is that the current version limits the number of players on a server to 64 or as few as 10, and a month after launch there were 129 instances operating in parallel, thus fragmenting the community. Another medium of communication is Twitch, the video service comparable to YouTube but focused on gaming. Players can stream a realtime video of building a colony in *ROK*H, and others can simultaneously contribute to a text forum seen on the same screen as the video, exchanging questions and answers about how to do this or that.

Third, as in more traditional alpha and beta tests, the *ROKH* designers can learn from the players. This may happen either explicitly, as when players send suggestions or criticisms to the designers, or implicitly, when the designers observe the behavior of players through the shared database that supports the virtual world. This has some similarity with how academics often use computer simulations, perhaps doing carefully designed experiments but certainly observing the consequences of assumptions built into the program.

Given that *ROKH* is very far from design completion, and changes are posted every few days, we cannot here perform a conclusive analysis, but an initial exploration can illuminate many implications of this unusual Mars colony simulation. On June 13, 2017, I created an avatar on a server with the ominous name, "Come die on Mars publicly," which could accept no more than 12 avatars and often had 2 or 3 online. One needs to name the avatar and select where to land, using a map divided into 100 squares, labeled A through J from west to east, and 1 through 10 from north to south. Only 10 of the squares were available, A1 through E2, and those south of this area were labelled "severe conditions: unsafe for landing." The eastern 50 squares were labeled "magnetic storm: not traversable," and I assumed this area really did not exist yet but was reserved for a later expansion of ROKH. In the western regions, the north was a starter area where one's avatar could prepare for an expedition into the dangerous south.

I selected region C2, and the loading screen carried this message: "Radiation is a constant threat on Mars. A cautious colonist will build their habitat out of the densest materials to block out as much cosmic rays as possible." My entry capsule landed me near a ruin from the earlier colonization attempt, so I explored it, and found no way to enter any of the half-wrecked buildings. The new, second colonization effort promised to send supplies, but they appeared almost at random in small containers scattered around the landscape. I soon found that they seemed to concentrate in and around the old outposts, some containing consumables like oxygen, food and water, while others contained construction components, such as tiles, spikes, rods, and raw

Fig. 4.4 A first, unsuccessful attempt to build a home on Mars

materials. In retrospect, I probably should have immediately starting building one of the homes called habitats, but I preferred to explore and to become familiar with the rather complex user interface of the software.

The formal measures of progress in *ROKH* are entirely based on one's growing collection of tools and raw materials. Starting out, one has an emergency screwdriver and an emergency hammer, both of which wear out quickly, but replacements are occasionally found in supply containers that drop from space. All the other tools must be constructed using these starter tools, or using more advanced tools in an ever more challenging if-then tree. The emergency screwdriver can make a pickaxe, using one medium rod and one small spike, which can be found occasionally in supply drums. The pickaxe can then be used to dig minerals, such as nickel and magnesium, the raw materials for making tiles that can be assembled into habitat houses, and components of various kinds of machinery. After wandering a while, I decided to settle down at the border between regions A2 and A3, not far from a wrecked colony, shown in Fig. 4.4.

On the left side of the picture is the very beginning of building a habitat by assembling tiles, 27 of which I either created with my emergency hammer after digging the materials with my pickaxe, or found in a supply drop. Clearly, many more tiles will be needed. The bright but tiny sun is shining just above the high point of the unfinished wall, and the wind-swept sands of the hill run across the bottom half of the image. In the background the many building of the ruined colony are visible, none of them accessible but valuable still because they are the target for the system that sends down supplies every once in a while.

The picture shows the user interface, not very visible in this publication, but providing an outline for discussion. The five squares at the bottom center, left to right, permit actions involving the screwdriver, the hammer, some food, the pickaxe, and some tiles that are in the inventory. The lower right corner of the display warns of problems, for example one of the tiny squares indicating that my avatar is carrying too much weight. There are five main variables, four of which being resources that must be kept up near 100%: energy, oxygen, food, and water which happens to be low at 52%. The fifth variable, radiation, measures damage, so it must be kept low, near 0%, but currently has risen to 46%. Hardly visible at all is a text chat in the lower left corner, which at the moment displays a message from a fellow player on this particular server, Aethir, who says: "Huh, lost power to work benchers…" This implies he had already built up a significant workshop, with advanced construction machinery that has run into some kind of bug.

Shortly after taking the screenshot in Fig. 4.4, I was forced to recognize that my avatar's radiation damage was increasing, and I did not seem to have any remedy for it. Some supply drops contain doses of potassium iodide, a chemical actually used as a treatment for radiation.[25] But in the area I was building the habitat, none of this *rad shield* could be found. Two solutions seemed plausible, but both would be costly. First, I created a second avatar on a different server in region B1, where the radiation tended to be low and I hoped that rad shield drops would be plentiful. Indeed, that turned out to be the case. Second, after operating the second avatar for a while, I returned to the first one, literally ran him rapidly to the same location where the second was prospering, and cured his radiation sickness with many doses of potassium iodide.

Figure 4.5 shows the second, much more successful attempt to build a habitat, by the second avatar. It is modest but very functional. At this point, it contained three storage chests I had made, one filled with surplus oxygen, and the other two with tools and raw materials. The door can be locked, so nobody can steal my property. The two solar energy devices on the roof are not yet connected to any machinery, but represent a start toward mass production. Several of the tools are required to make each other as well as advanced equipment, but one is a portable electrolyzer that can purify water mined from ice using a pickaxe, so it becomes drinkable. The small box in the lower right corner is one of the supply drops, and in the background behind it are the Apollo-like landing capsule the avatar arrived in, and a ruined interplanetary communication dish antenna.

Each of the two solar energy devices is a medium-sized solar plate, requiring a rivet gun to assemble two solar panels, a medium metal sheet and a medium rod. Each solar panel requires a welder machine to assemble a small metal or glass sheet, a small insulator, and four wires. Making the welder required the rivet gun to assemble a 1H case, a small tank, a small compressor, a medium tube and a coil. A rivet gun can be made using an assembly tool and a small rod, a small axle, a coil and a small spike. The small axle is made from a small rod using a utility knife. A coil can be made from 6 units of mineral, by means of a manufacturing tool. The assembly tool is made from a small rod and small spike using an emergency screwdriver. The compressor requires use of a portable drill on a small sphere, which was made from 22 units of

[25]https://en.wikipedia.org/wiki/Potassium_iodide, accessed December 2017.

Fig. 4.5 A second, successful attempt to build a home on Mars

mineral using the manufacturing tool, which itself was made with the assembly tool from a small rod and a small bar. Confusing? Yes, but also educational, not because these procedures would actually work in the real world, but because they illustrate the abstract principles of if-then logic which govern computer science as well as the economics of manufacturing here on Earth.

The official *Field Operations Manual* for *ROKH* does not explain clearly how to perform each action, which one must deduce from experimenting with the tools and raw materials or learn from other players. Rather, it illustrates the range of goals one may seek to achieve:

> Beyond a simple habitat for shelter, designing your own structures can solidify your control and give you more options:
>
> A pressurized environment where it's possible to open your helmet safely
>
> Workbenches which provide a huge increase to your crafting capabilities
>
> Chests and storage for surplus goods and materials
>
> Automated survival equipment which collect oxygen and power continuously
>
> Customize your habitat with multiple wings, floors, lights, and windows. Use tiles with different materials for a unique look.
>
> Keep yourself and your possessions safe from other colonists by locking your doors. Only those on the Permissions List will then be able to have access[26]

Indeed, before I had completed the habitat in Fig. 4.5, complete with a lockable door, supplies had vanished from a storage tank where I had put vast amounts of oxygen and from a chest containing raw material for manufacturing. Thieves! I did encounter three different fellow players on the particular server, but its typical population was never more than 3 when I logged in, and usually 1 or 0. There were many fragmentary structures strewn across the landscape, indicating that several people had visited Mars only briefly, or had been checking out multiple servers.

[26]https://steamcommunity.com/sharedfiles/filedetails/?id=923530815.

My neighbors had constructed two advanced manufacturing machines and placed them outdoors so any of us could use them: a metal press that could make improved versions of many of the components such as rods and spikes, and a tile mold for mass production of architectural elements.

One of the more social dimensions of *ROKH* is the series of often rather long playthroughs a few people have posted on the game-oriented competitor to YouTube, named Twitch. On June 12, 2017, MJ Guthrie streamed in realtime her first hour and 17 min exploring *ROKH*, intentionally having essentially no preparation, so we could watch her figure it out by observing, experimenting, and thinking: "MJ finally gets to live out a dream of living on Mars. Of course, how long she actually lives is questionable in this new survival game."[27] She did well, because she is an expert game blogger with a background in psychology and a fascination with the dynamics of virtual communities.[28] The way Twitch streaming works, people watching can type in text comments, which will be saved and displayed at the appropriate times to anyone who watches the video in later days. Here, for a very real but somewhat obscure example, are some comments people made as MJ was learning how to understand *ROKH* when meteors started crashing down:

ThBeatnik : Yes, MJ is absolutely ninja-ing another player's phat 100t!!!!

dentadsio : yo

ThBeatnik : Supply drops?

natalyiatsw : Nuke them from orbit - it's the only way to be sure…

ThBeatnik : *watches the O2 read-out; frets*

BalsBigBrother : Try to catch one. What's the worst that could happen :p

BenSilencing : MJ, did you hear about Ylands? Announcement reminded my of Everquest Next in some ways.

ThBeatnik : ((That might have been a "get out of the way" UI thingy))

KatrianaND : floaty burns the fuel FAST it looks like

ThBeatnik : Looks like it needs a battery?

KatrianaND : it's says charge battery, not change battery

ThBeatnik : Oho! It somehow recharged your power level.

Bauvil : but what about o2

ThBeatnik : Whew.

Among the several other *ROKH* videos on Twitch, one is an hour-long interview with its developers, by the Gamasutra game news organization.[29] The online Reddit forums have a *ROKH* section, and there is a small Facebook group for *ROKH* as well.[30] By Saturday, July 15, 2017, two months after early access release, *ROKH* had not attracted many players. A total of 253 people had posted casual reviews on

---

[27] www.twitch.tv/videos/152039200, accessed December 2017.

[28] https://massivelyop.com/author/mj-guthrie, accessed December 2017.

[29] www.twitch.tv/videos/149035853

[30] www.reddit.com/r/RokhTheGame; www.facebook.com/groups/1884519868504204, accessed December 2017.

Steam, just 144 of them or 57% positive. At 4:30 in the afternoon, Eastern US time, the total number of players online was just 41, 18 of them on one of the locked private servers, and the remainder distributed across 17 other servers.

4.4 Where Education May Lead

A lingering question about educational simulations concerns whether the factors that elicit the interest of users are identical with the factors that communicate correct information. A rather famous example, in which this issue expanded into a public controversy, was the 2008 game, *Spore*. It models the evolution of biological intelligence up through the stage of interstellar colonization, and as *Science* magazine journalist John Bohannan reported, had educational pretensions: "The game's makers are clearly aiming for the highly lucrative family and education markets. 'Since the game's release we've received a lot of interest from various schools and universities around the world,' a *Spore* spokesperson wrote me in an e-mail. 'So that's a good sign that there's a lot of interest in [the] academic/education community (Bohannon 2017).'" Further evidence of the pretension to be educational is the fact that Spore was publicized through a *National Geographic* television documentary that found some scientific merit in it.[31] *Spore's* own website describes its cosmic ambitions:

> With Spore you can nurture your creature through five stages of evolution: Cell, Creature, Tribe, Civilization, and Space. Or if you prefer, spend as much time as you like making creatures, vehicles, buildings and spaceships with Spore's unique Creator tools.
>
> CREATE Your Universe from Microscopic to Macrocosmic - From tide pool amoebas to thriving civilizations to intergalactic starships, everything is in your hands.
>
> EVOLVE Your Creature through Five Stages - It's survival of the funniest as your choices reverberate through generations and ultimately decide the fate of your civilization.
>
> EXPLORE your world and beyond - Will you rule, or will your beloved planet be blasted to smithereens by a superior alien race?
>
> SHARE with the world - Everything you make is shared with other players and vice versa, providing tons of cool creatures to meet and new places to visit.
>
> While Spore is a single player game, your creations and other players' creations are automatically shared between your galaxy and theirs, providing a limitless number of worlds to explore and play within.[32]

Note the technical detail that *Spore* is a solo-player game that permits a degree of communication between players, a quality shared with *No Man's Sky*, considered here in Chap. 6. And it is highly social in the same way as the simulations described in Chap. 2, and the games in Chap. 5, because the user interacts with a simulated community of other persons and creatures. They may possess a degree of artificial intelligence, but they definitely interact socially with each other and with the user.

---

[31] https://shop.nationalgeographic.com/product/dvds/science/how-to-build-a-better-being-dvd-exclusive, accessed June 2017.

[32] www.spore.com/what/spore, accessed June 2017.

At times, the user is represented by a specific creature inside *Spore*, effectively the user's avatar, in which case it resembles a role-playing game. But often the user is not represented inside the simulation, but operates at a distance. This genre is sometimes called *god games*, and Wikipedia references *Spore* in the page devoted to this term: "A god game is an artificial life game that casts the player in the position of controlling the game on a large scale, as an entity with divine and supernatural powers, as a great leader, or with no specified character (as in *Spore*), and places them in charge of a game setting containing autonomous characters to guard and influence."[33]

The original video game trailer for *Spore* displays vigorous cartoonish images of evolving lifeforms that begin as single-cell organisms in the ocean, come up upon land, evolve society and civilization, eventually colonizing the galaxy.[34] The narration begins: "Behold the galaxy, full of life and stars and intelligence! But it wasn't always like this. Not long ago it was cold and dark. So what happened? Someone made a decision. Someone looked at a microbe and thought, 'He could use a longer flagella, and a pair of pincers, too.'" This unnamed *someone* keeps making decisions, improving the lifeform until it gains intelligence and ultimately flies to the stars. Was it God who made all these decisions of biological evolution, bringing light to the darkness? The narration ends: "But what they never realized, was that all along the way, from humble microbes to starship captains, someone had guided them at every turn. And that someone is you." So, *Spore* puts the user in the role of God, applying a theory of intelligent design to evolution, rather than natural selection from random variation.

Considered from the standpoint of popular videogames, and indeed most other forms of popular culture, *Spore* is quite normal when it imposes human meaning upon natural phenomena. In his contribution to a collection of cognitive science essays, Porter Abbott suggests that human thought organizes things in terms of narratives - stories in which protagonists face obstacles and take actions in pursuit of goals - and thus the scientific theory of evolution is *unnarratable* (Porter Abbott 2003). Yes, we can state evolutionary theory in terms of algorithms incorporating random number generators that model mutation, then population dynamics in which factors in a complex system impose natural selection, and well-educated people may understand the principles involved. But this is not a story that can be the basis for a fully engaging game, or indeed a novel or grand opera.

As a medium for communicating scientific theories and empirical truths, however, the narrative of an engaging game may seriously distort nature even as it attracts human interest. Wikipedia reports: "In October 2008, John Bohannon of *Science* magazine assembled a team to review the game's portrayal of evolution and other scientific concepts.[35] Evolutionary biologists T. Ryan Gregory of the University of Guelph and Niles Eldredge of the American Museum of Natural History reviewed the Cell and Creature stages. William Sims Bainbridge, a sociologist from the U.S.

---

[33]https://en.wikipedia.org/wiki/God_game, accessed June 2017.

[34]www.youtube.com/watch?v=SUFLou_d4uw, accessed June 2017.

[35]Bohannon, John. 2008. Flunking Spore. *Science,* pp. 322

Fig. 4.6 Creatures creating cross-species friendship by singing

National Science Foundation, reviewed the Tribe and Civilization stages. NASA's Miles Smith reviewed the Space Stage. The *Science* team evaluated Spore on twenty-two subjects. The game's grades ranged from a single A in galactic structure and a B + in sociology to Fs in mutation, sexual selection, natural selection, genetics, and genetic drift."[36]

Because it is a *strategy game*, *Spore* may encourage the development of young minds, by requiring solution of abstract problems within its narrative, however. For example, Fig. 4.6 is a scene in the Creature chapter, when the player does have an avatar, and mine is the animal in the foreground. He has encountered a half dozen animals of a different species, and wants to make peace with them rather than cause conflict in which he would be outnumbered. So he sings to them, and they sing back. This does not work with all the other species, and the alternate strategy is for my avatar to breed with others of his species - operated by simple artificial intelligence rather than being avatars of other players - to build a large population that can defeat any species with which it does not build a friendship.

Bohannon's team of scientific critics published a book chapter, analyzing *Spore's* strengths and weaknesses in some detail (Bohannon et al. 2010). My evaluation of the Tribe and Civilization stages was more positive than my co-authors' evaluations of the stages of biological evolution, so here I will summarize my observations. The most

[36]https://en.wikipedia.org/wiki/Spore_(2008_video_game), accessed June 2017.

appropriate perspective was to consider how well *Spore* harmonized with theories of cultural evolution, however poorly it did with the science of biological evolution. To be sure, there is less scientific consensus concerning the cultural sciences, compared with the natural sciences, but that implies that the primary responsibility for educators becomes communicating ideas more than facts, using facts primary to illustrate ideas and to serve as segues to other ideas when the facts seem to contradict a theory.

Spore's division of history into five eras is not unreasonable. Since ancient days, intellectuals have tended to conceptualize history in terms of a series of distinguishable eras of time, such as the period when Adam and Eve lived in the Garden of Eden, the history from their expulsion until Moses brought the Ten Commandments down from the mountaintop, the period until the ministry of Jesus, and the following Christian Era. Anthropologist Claude Lévi-Strauss reported that many pre-literate people had mythologies dividing time into two periods, which he called *the raw and the cooked*, before and after the emergence of the particular culture (Lévi-Strauss 1970). V. Gordon Childe has argued that the key watershed in human history was the Neolithic Revolution around ten thousand years ago, in which the domestication of plants and animals led to social systems that enforced ownership of agricultural land, the development of complex economies based on division of labor, and the establishment of the first cities in a context of growing population (Gordon Childe 1951). Many authors conceptualized the development of steam engines and mass production in factories as the Industrial Revolution, leading a recent encyclopedia of world history to distinguish three main phases: the Era of Foragers, the Agrarian Era, and the Modern Era (McNeill 2005).

In geology and paleontology, a universal system for dating rocks and fossils is maintained by the International Commission on Stratigraphy that proclaims its "primary objective is to precisely define global units (systems, series, and stages) of the International Chronostratigraphic Chart that, in turn, are the basis for the units (periods, epochs, and age) of the International Geologic Time Scale; thus setting global standards for the fundamental scale for expressing the history of the Earth."[37] The smaller types of period vary somewhat across geographic regions, but the overall system is regarded as a highly valuable and rather accurate dating system, given that quantitative methods such as carbon isotope dating are only rarely available. A key factor providing empirical support for the geological periods is the emergence and diffusion of new lifeforms, preserved as fossils, which may reflect major environmental changes as well as simply the long-term emergence of greater complexity in lifeforms or temporary prominence of trilobites and dinosaurs.

Given the antiquity of notions that humanity has experienced a sequence of historical eras, a framework already existed for social theorists of the nineteenth century to begin to transfer evolutionary concepts between biology and sociology. Herbert Spencer was a leader in both fields, applying similar concepts to them, although in a rather more sophisticated manner than the catch phrase "survival of the fittest" that he coined. He believed that each cause in our universe has more than one effect, so all forms of long-term evolution would lead to increasing complexity (Spencer 1857).

[37]www.stratigraphy.org, accessed June 2017.

In the 1970s a school of thought often called *sociobiology* emerged, and its proponents struggled to develop a comprehensive theory to explain both biological and cultural evolution among humans (Wilson 1975; Van den Berghe 1975; Dawkins 1976; Lumsden and Wilson 1981; Luca Cavalli-Sforza and Feldman 1981; Bainbridge 1985). However, a crucial variable in both biological and social evolution was missing from *Spore*, namely the family.

As its instruction manual explains, the Tribe phase in *Spore* is organized around "a small, close-knit village community, complete with a totem and a central Tribal Hut (Hodgson 2008)." Yet tribes were extended families in the tribal phase of real world social evolution (Lévi-Strauss 1969). From the Tribe stage onward, *Spore* resembles the strategy games described in the following chapter of this book, and the Civilization stage involves building cities and deploying increasingly advanced technologies, until the player is ready for the Space stage. Three alternative strategies are available in the Civilization stage, each emphasizing one societal institution: military, religion, or economy. The manual defines them thus: "A military nation seeks to expand its borders by conquering all other cities, while a religious society wins control of other cities by winning the hearts of its citizens. An economic culture creates vast wealth and uses it to expand its sphere of influence (Hodgson et al. 2008)." These three factors also operate in the Tribe phase, and economic exchange was certainly also a factor in the real world, becoming more fully standardized with the growth of complex civilizations (Malinowski 1922; Polanyi 1944; Mauss 2000).

One way a player may conquer a city in the Civilization phase is to undercut its economy, then start flooding its residents with religious propaganda that causes them to become disloyal. This combination of two strategies is sophisticated, but only partly supported by the social science of religion. Indeed, religious appeals work best with deprived populations either because religion is fundamentally a psychological compensator against the deprivations of human life, or because they have no alternative in their desperation (Pope 1942; Cohn 1961; Smelser 1962; Stark and Bainbridge 1987). However, religious conversion does not take place through impersonal *disembodied appeals*, but through intimate social networks, and religions are best conceptualized as expressions of existing social systems (Durkheim 1915; Lofland and Stark 1965; Shupe 1976; Stark and Bainbridge 1980).

While *Spore* drew much academic criticism for apparently promoting a theistic *intelligent design* theory of biological evolution, the same criticism could be applied to the technological evolution that enabled the Space stage. A school of sociological thought called *technological determinism* argues that technological innovation is the autonomous engine of social change, and that innovation itself is self-generating (Ogburn 1922; White 1959). Since the original industrial revolution, social scientists have sought to understand and even control the human consequences of technical change, yet this is very different from putting one person in control of human history (Smith 1812; Owen 1813). Figure 4.7 shows the glorious, if improbable, result of *Spore* simulation of society, the launch of a spaceship looking rather like a flying saucer, from a virtual city, amidst a brilliant display of propulsive fireworks.

Fig. 4.7 The launch that transitions from civilization to space

According to a *Spore* wiki, the Space stage "deals with various scientific and science fictional concepts, such as colonization, astrobiology, interaction with alien races, the galactic topography (how the astronomical phenomena and planets interact), terraforming and various missions."[38] The in-game instruction manual described the Space Stage as liberation to seek one's own goals:

> Will you expand your empire through terraforming and colonizing planets in outlying star systems?
>
> Will you discover other races and befriend them to form an intergalactic federation or throw down the gauntlet and initiate intergalactic wars?
>
> Will you explore the farthest reaches of the galaxy in search of riches?
>
> Will you become the galaxy's most renowned trader?
>
> Will you perhaps seek out deeper meaning in a quest to the center of the galaxy?

The printed *Spore* instruction manual explains that achievement of each of these goals may require obsessive gathering of a valuable natural resource called *spice*: "Spice is the most sought-after resource in the galaxy. It powers cities, colonies, vehicles, and ships. It's responsible for making poor empires rich, corrupting the innocent, and sparking interstellar wars. Most will do whatever it takes to acquire it and pay big money to whomever has it (Anonymous 2008a)." It is worth noting

[38] https://spore.wikia.com/wiki/Spore, accessed June 2017.

that one builds a galactic economy based on spice by exploring out into the universe and colonizing planets. This contrasts with a more conservative model of interstellar colonization that would require the colonies to send profits back to Earth, to pay the debt of building the technology with which to colonize. Colonizing planets is essential in the *Spore* universe:

> To become the dominant species, you must build a network of colonies that stretch across the galaxy. Colonies perform the following functions:
>
> Generate Income: Colonies can be plentiful sources of spice, which is turned into Sporebucks that you can spend on building your empire.
>
> Repair and Refuel: You can repair your damaged spaceship and replenish its energy at any colony that you control.
>
> Colony Vehicles: As your empire spreads across the galaxy, you may gain control of additional vehicles. As you improve and develop your colonies, colony vehicles are automatically produced. They are a good indicator as to how well your colonies are doing.
>
> When looking for planets to colonize, you should aim for planets that have a strategic location or are good sources of spice. Targets should be habitable with little or no small terraforming investment (Anonymous 2008b).

Spore raises very legitimate questions about whether educational computer games need to be delivery vehicles for regular academic curriculum, versus environments where designers may express philosophical values and players may learn innovative ways of thinking and acting. Professional scientists and other academics may favor "accuracy" but companies and their customers may prefer engaging stories.

4.5 Conclusion

Frankly, I find the current state of educational spaceflight software rather shocking, considering that the high quality of expertise simulation in *Buzz Aldrin's Space Program Manager* and environment simulation in *ROKH* are not repaid by popularity. It is one thing to note that the popularity of science fiction literature based on real phenomena in the natural sciences has declined greatly in recent decades, losing ground to fantasy which it had dominated in the middle of the twentieth century (Benford 2015; Bainbridge 1986). It is quite another to contemplate the possibility that the most science-based new media online are failing to meet our expectation for generating public enthusiasm and popular awareness of the real possibilities. Perhaps close examination of more popular space-related games in the following chapters will justify greater optimism.

Chapter 1 already introduced two virtual universes that could serve educational functions, *Second Life* and *Star Trek Online*. Around 2008, when *Second Life* was at its greatest popularity, innovative educators flocked to it, setting up virtual college campuses where students could learn either computer and information science, or other disciplines that could effectively be represented in this online environment. Remarkably, *Star Trek Online* includes a system called the Foundry, that allows players to create their own instanced missions that other players can experience and evaluate.[39] Perhaps educational researchers should reverse their approach, not creating games as instruments for delivering a traditional curriculum, but seeking ways through which students could create their own games. The first wave of online educational game activity sought merely to produce prototypes, but the next wave will need to give greater emphasis to understanding, conducting research not merely to determine the most effective methods of teaching, but also to examine the mental processes and personal motivations possessed by the students.

Chapter 2 illustrated how the colonization of Mars could be used as a framework for simulating standard theories in social science, with initially educational goals. From the standpoint of mathematics, statistics, and computer science, the best modality for educational simulations would be abstract, involving no particular narrative. Yet humans tend to like activities that have a background story, and some social scientists might argue that the proper lore for an educational simulation would be historical. In fact, there do exist a few historical massively multiplayer games that could be used to teach principles of sociology, economics or political science. Two notable examples are *A Tale in the Desert* that emulates ancient Egypt and *Pirates of the Burning Sea* that does so for the Caribbean in 1720, yet both are rather unpopular. Science fiction is somewhat more popular, and encourages admiration of real science, so its lore would be more appropriate. This leaves open the question of how realistic in terms of physical science the simulations should be, if their primary goal is educational representation of social science.

References

Abbott,H. Porter. 2003. Unnarratable Knowledge: The Difficulty of Understanding Evolution by Natural Selection. In *Narrative Theory and the Cognitive Sciences*, ed. David Herman, pp. 143–162. Stanford, California: Center for the Study of Language and Information.

Anonymous 2008a. *Spore Instruction Manual* Redwood City, California: Electronic Arts, p. 51.

Anonymous 2008b. *Spore Instruction Manual* Redwood City, California: Electronic Arts, p. 50.

Asbell-Clarke, Jodi, Teon Edwards, Elizabeth Rowe, Jamie Larsen, Elisabeth Sylvan, and Jim Hewitt.1989. Martian Boneyards: Scientific Inquiry in an MMO Game. *International Journal of Game-Based Learning*, 2(1), 52–76, p. 52.

Avery, Phillipa and Sushil Louis. 2010. Coevolving Team Tactics for a Real-Time Strategy Game. In *Proceedings of the 2010 IEEE Congress on Evolutionary Computation*, pp. 1–8. Piscataway, New Jersey: IEEE.

Bainbridge, William Sims. 1985. Cultural Genetics. In *Religious Movements*, ed. Rodney Stark, 157–198. New York: Paragon.

Bainbridge, William Sims. 2015. *The Meaning and Value of Spaceflight*, 184–185. Berlin: Springer.

Bainbridge, William Sims. 1986. *Dimensions of Science Fiction*. Cambridge: Harvard University Press.

Ballinger, Christopher and Sushil Louis. 2013. Comparing Coevolution, Genetic Algorithms, and Hill-Climbers for Finding Real-Time Strategy Game Plans. In *Proceedings of the 15th Annual Conference Companion on Genetic and Evolutionary Computation*, p. 1. New York: ACM.

[39]https://sto.gamepedia.com/Guide:_The_Foundry, accessed December 28, 2017.

Barab, Sasha, Michael Thomas, Tyler Dodge, Robert Carteaux and Hakan Tuzun. 2005. Making learning fun: quest atlantis, a game without guns. *Educational Technology Research and Development* 53(1):86–107, p. 95.
Barab, Sasha A., Melissa Gresalfi, and Adam Ingram-Goble. 2010. Transformational Play: Using Games to Position Person, Content, and Context. *Educational Researcher* 39(7): 525–536, p. 526.
Benford, Gregory. 2015. The State of the SF Magazines. www.gregorybenford.com/uncategorized/the-state-of-the-sf-magazines.
Bohannon, John. 2008. Flunking Spore. *Science*, pp. 322.
Bohannon, John, T. Ryan Gregory, Niles Eldredge, and William Sims Bainbridge. 2010. Spore: Assessment of the Science in an Evolution-Oriented Game. In *Online Worlds: Convergence of the Real and the Virtual*, ed. William Sims Bainbridge, pp. 71–86. London: Springer.
Bronner, Fritz. 1992. *Buzz Aldrin's Race Into Space: Rules of Play*, 1. Irvine, California: Interplay.
Cavalli-Sforza, Luigi Luca and Marcus W. Feldman. 1981. *Cultural Transmission and Evolution.* Princeton, New Jersey: Princeton University Press.
Childers, Richard. 2010. A Virtual Mars. In *Online Worlds: Convergence of the Real and the Virtual* edited by William Sims Bainbridge, pp. 101–109 London: Springer, p. 108.
Cohn, Norman. 1961. *The Pursuit of the Millennium*. New York: Harper.
Dawkins, Richard. 1976. *The Selfish Gene*. New York: Oxford University Press.
Durkheim, Emile. 1915. *The Elementary Forms of the Religious Life*. London: Allen and Unwin.
Garfinkel, Harold. 1967. *Studies in Ethnomethodology*. Englewood Cliffs, New Jersey: Prentice-Hall.
Gordon, V. 1951. *Childe, Man Makes Himself*. New York: New American, Library.
Hodgson, David S.J., Bryan Stratton, and Michael Knight. 2008a. *Spore: Official Game Guide*, 87. Roseville, California: Prima Games.
Hodgson, David S.J., Bryan Stratton, and Michael Knight. 2008b. *Spore: Official Game Guide*, 119. Roseville, California: Prima Games.
Lévi-Strauss, Claude. 1969. *The Elementary Structures of Kinship*. Boston: Beacon Press.
Lévi-Strauss, Claude. 1970. *The Raw and the Cooked*. New York: Harper.
Liu, Siming, Sushil J. Louis and Monica Nicolescu. 2013. Comparing Heuristic Search Methods for Finding Effective Group Behaviors in RTS Game, In *Proceedings of the 2013 IEEE Congress on Evolutionary Computation,*pp. 1371–1378. Piscataway, New Jersey: IEEE.
Lofland, John, and Rodney Stark. 1965. Becoming a World-Saver: A Theory of Conversion to a Deviant Perspective. *American Sociological Review* 30: 862–875.
Lumsden, Charles J., and Edward O. Wilson. 1981. *Genes, Mind, and Culture*. Cambridge, Massachusetts: Harvard University Press.
Malinowski, Bronislaw. 1922. *Argonauts of the Western Pacific*. London: Routledge.
Mauss, Marcel. 2000. *The Gift*. New York: Norton.
McNeill, William H. (ed.). 2005. *Berkshire Encyclopedia of World History (Great Barrington.* Berkshire: Massachusetts.
Ogburn, William F. 1922. *Social Change with Respect to Culture and Original Nature.* New York: B. W. Huebsch.
Owen, Robert. 1813. *A New View of Society*. London: Cadell, and Davies.
Polanyi, Karl. 1944. *The Great Transformation*. New York: Farrar and Rinehart.
Pope, Liston. 1942. *Millhands and Preachers.* New Haven, Connecticut: Yale University Press.
Sandvig, Christian, Kevin Hamilton, Karrie Karahalios, and Cedric Langbort. 2016. When the Algorithm Itself Is a Racist: Diagnosing Ethical Harm in the Basic Components of Software. *International Journal of Communication* 10: 4972–4990.
Schutz, Alfred. 1967. *The Phenomenology of the Social World*. Evanston, Illinois: Northwestern University Press.
Shupe, Anson D. 1976. 'Disembodied Access' and Technological Constraints on Organizational Development. *Journal for the Scientific Study of Religion* 15: 177–185.
Smelser, Neil J. 1962. *Theory of Collective Behavior*. New York: Free Press.

Smith, Adam. 1812. *An Inquiry into the Nature and Causes of the Wealth of Nations*. London.
Smith, Peter A. and Alicia Sanchez. 2010. Mini-Games with Major Impacts, pp. 1–12 in *Serious Game Design and Development: Technologies for Training and Learning*, ed. Janis Cannon-Bowers and Clint Bowers, p. 6. Hershey, Pennsylvania: IGI Global.
Spencer, Herbert. 1857. Progress: Its Law and Causes. *The Westminster Review* 67: 445–485.
Stark, Rodney, and William Sims Bainbridge. 1980. Networks of Faith. *American Journal of Sociology* 86: 1376–1395.
Stark, Rodney, and William Sims Bainbridge. 1987. *A Theory of Religion*. New York: Toronto/Lang.
Van den Pierre, L. 1975. *Berghe, Man in Society: A Biosocial View*. New York: Elsevier.
Vertesi, Janet. 2014a. *Seeing Like a Rover: Images in Interaction on the Mars Exploration Rover Mission*. Chicago, Illinois: University of Chicago Press.
Vertesi, Janet. 2014b. *Seeing Like a Rover: Images in Interaction on the Mars Exploration Rover Mission*, 8. Chicago, Illinois: University of Chicago Press.
White, Leslie A. 1959. *The Evolution of Culture*. New York: McGraw-Hill.
Wilson, Edward O. 1975. *Sociobiology: The New Synthesis*. Cambridge, Massachusetts: Harvard University Press.

Chapter 5
Computer Simulation for Space-Oriented Strategic Thinking

Although typically presented as playthings or recreational puzzles, strategy games are highly intellectual, requiring careful analysis, calculating what resources to obtain in what order, and how to invest these resources in further expansion beyond the starting point. Many of the highest quality and most popular computer-based examples are wargames, often in historical settings, and the Buzz Aldrin examples covered in the previous chapter simulated space-related projects during the Cold War. This chapter will consider a range of strategy games set largely on other planets, some of which imitate interstellar colonization, albeit typically in the context of conflict with alien civilizations. The specific details of these examples may not model our real space future particularly well, but all are excellent mental exercise, requiring the user to think deeply about abstract problems that may turn out more relevant to the real future than their particular instantiation in the software.

5.1 The Most Influential Fictional Mars

Modern computer-based strategy games draw upon multiple historical traditions, including table-top military war games, as Wikipedia reports: "Manual simulations have probably been in use in some form since mankind first went to war. Chess can be regarded as a form of military simulation (although its precise origins are debated). In more recent times, the forerunner of modern simulations was the Prussian game *Kriegsspiel*, which appeared around 1811 and is sometimes credited with the Prussian victory in the Franco-Prussian War. It was distributed to each Prussian regiment and they were ordered to play it regularly, prompting a visiting German officer to declare in 1824, 'It's not a game at all! It's training for war!'[1] The reference to chess provides a link to outer space through a century-old variant of chess set on the planet Mars, called *jetan*.[2]

[1] en.wikipedia.org/wiki/Military_simulation, accessed December 2017.

[2] en.wikipedia.org/wiki/Jetan, accessed December 2017.

W. S. Bainbridge, *Computer Simulations of Space Societies*, Space and Society,
https://doi.org/10.1007/978-3-319-90560-0_5

Devised by science fiction writer Edgar Rice Burroughs who frequently played it for his own amusement, jetan was presented to the public in his 1922 novel *Chessmen of Mars*. The game played an important role in the story, and an appendix documented the rules. Like chess, it was played on a physical board composed of squares, but 10 by 10 rather than 8 by 8, presumably because the decimal system seemed more scientific in an era before 8-bit bytes became standard in computing. Jetan is not only a good first example for this chapter of a classical strategy game, but also connects to computer simulations in two ways. First, it was among the inspirations for the original *Dungeons and Dragons* (D&D) table-top game that powerfully influenced the development of massively multiplayer online role-playing games, like not only *Dungeons and Dragons Online* and *Neverwinter* that are direct expression of D&D, but also *World of Warcraft* (WoW) and many examples described in later chapters. Indeed, the original 1994 WoW version, *Warcraft: Orcs & Humans*, is based on a story similar to that of jetan (Bainbridge 2010). Second, there have been a couple of cases in which amateur fans of Burroughs have created computerized versions of jetan.

Chessmen of Mars is the fifth in a series of novels Burroughs wrote about imaginary, competing civilizations on Mars. The first, *A Princess of Mars*, written in 1911, follows an American named John Carter who is psychically transported to the red planet where he experiences intense adventures interacting with two mutually hostile intelligent humanoid species, the Green Martians and the Red Martians. The Green Martians are like the Orcs in WoW, violent tribal warriors, while the Red Martian are civilized people more like Earthlings. Carter falls in love with Dejah Thoris, the eponymous princess of Mars, a Red Martian, but develops a lasting friendship with Tars Tarkas, a Green Martian. Arguably, Tarkas was the first extraterrestrial character very different from humans who was given a complex personality and culture in science fiction. In addition, the series of Mars novels by Burroughs were the direct inspiration for much later science fiction, especially the *Star Wars* mythos that features here in Chap. 8 (Bainbridge 2016). As the instructions in *Chessmen of Mars* explain, jetan simulates an ancient enmity between two other Martian races:

> The game is played with twenty black pieces by one player and twenty orange by his opponent, and is presumed to have originally represented a battle between the Black race of the south and the Yellow race of the north. On Mars the board is usually arranged so that the Black pieces are played from the south and the Orange from the north (Burroughs 1922).

As in chess, at the beginning, each player has a set of varied pieces arranged in two lines on the nearest squares of the board. Like pawns, the further line of squares is filled with 8 *panthans*. In the glossary appendix to his previous Mars novel, Burroughs had defined *panthan* as "soldier of fortune (Burroughs 1920)." Flanking them, to complete a row of 10 pieces, are two *thoats*, pieces with a move comparable to that of knights in chess. Technically, in the Martian language invented by Burroughs, thoats are the monstrous animals the knights ride:

> A green Martian horse. Ten feet high at the shoulder, with four legs on either side; a broad, flat tail, larger at the tip than at the root which it holds straight out behind while running; a mouth splitting its head from snout to the long, massive neck. It is entirely devoid of hair

> and is of a dark slate color and exceedingly smooth and glossy. It has a white belly and the legs are shaded from slate at the shoulders and hips to a vivid yellow at the feet. The feet are heavily padded and nailless (Burroughs 1920).

At the center of the row of squares nearest the player are pieces comparable to the king and queen of chess, the chief and princess. Flanking them to left and right are a flier, dwar (captain), padwar (lieutenant), and warrior, with moves more complex than in chess, because they can change direction, but limited in how many spaces they can move each turn. For example, a flier can go "3 spaces diagonal in any direction or combination; and may jump intervening pieces (Burroughs 1922)."

Players take turns moving one piece at a time, as in chess, and one way to win is for one's chief to kill the opponent's chief, by landing on the same square. Landing any piece other than one's chief in the same square as the opponent's chief does not win the game, but causes an immediate draw, with no winner. Thus, a chief in jetan is far more aggressive than a king in chess, who tends to hide like a coward behind his minions. A chief moves three spaces on a turn, in any combination of directions, but cannot fly over intervening pieces as a flier can.

A second way to win is to get any one of one's pieces onto the same square as the opponent's princess. This is conceptualized as capturing her, rather than killing her. She has the same moves as her chief, except that like a flier she can jump over intervening pieces, and: "The Princess may not move onto a threatened square, nor may she take an opposing piece. She is entitled to one ten-space move at any time during the game. This move is called *the escape* (Burroughs 1922). Clearly, the fundamental logic of this game follows the narrative structure of many of the stories Burroughs wrote, which involve rescuing non-violent princesses like Dejah Thoris, and require the protagonist to be heroic. Googling "jetan" turns up a large number of websites describing the game, starting with its Wikipedia page that describes how it featured in the story of the novel:

> The second half of *The Chessmen of Mars* takes place in the city of Manator, where the most popular civil event involves human beings fighting to the death in a life-sized Jetan game viewed by hundreds of spectators. The "board" is large enough that some of the pieces are mounted on Thoats and yet still fit in a single "square". However, this life-and-death version departs from the rules of Jetan in one very significant way: When one piece lands on a square occupied by another, the first does not automatically replace the second. Rather, the two pieces fight to the death, and the winner of the sword fight wins the square. The lone exception involves the Princess: if one side's piece lands on a square occupied by the other side's Princess, no battle occurs, and the first side wins the game.[3]

An online version of jetan created in 2004 by "Mad Elf" and still available in 2017, is very much a simulation, because it allows some changes to the rules of the game, and thus performing experiments with the software.[4] For example, the rule selection screen allows a player to follow Burroughs' move rule for the Princess, or alter it in one of several ways:

---

[3]en.wikipedia.org/wiki/Jetan, accessed May 2017.

[4]www.maranelda.org/extras/jetan/help.html, accessed May 2017.

Moves three squares, can jump (Can't take pieces or move to threatened squares)
Must move exactly three squares (chained)
Can move one, two or three squares (free)
Can move orthogonally or diagonally, but can't change direction
Can move orthogonally or diagonally, but not both in the same move (civil)
Can move orthogonally and diagonally in the same move (wild)
Can move across threatened squares (brave)
Can't move across threatened squares (frightened)[5]

Another computer version of jetan, programmed by Paul Burgess in 1993, is downloadable software for playing jetan against one's computer.[6] The zip file includes the source code of the program, which was written in the excellent but now unfashionable language Pascal. For example, here is a procedure that need not be fully analyzed, but is rather self-explanatory:

```
function incheck (var list : movelist; pl : player; var bd : board) : boolean;
{ Find whether princess is in check }
var
i : integer;
princess : squarenum;
begin
with list do begin
i := nmoves;
with bd do
princess := chessmen[pl,numchessmen[pl]];
if bd.sq[princess] in [oprincess,bprincess] then begin
moveto[0] := princess;
while moveto[i] <> princess do
i := i - 1;
incheck := i > 0
end
else { princess has already been taken }
incheck := true
{ incheck := (i > 0) or not (bd.sq[princess] in [oprincess,bprincess]) }
end
end; { incheck }
```

[5] www.maranelda.org/extras/jetan/setup.html, accessed May 2017.
[6] pmburgess.blogspot.com/2005_12_01_pmburgess_archive.html, accessed May 2017.

Recent strategy games tend to build on the principles of jetan, especially as it was played in Manator, by imagining a much more complex environment than a 10 by 10 board of squares. Many build upon a map of many equal-size hexagons rather than a few squares, but then place various barriers and resources upon particular hexagons. In standard gamer terminology, the best strategy games are *4X*. This is an obscene pun meaning sexy, but was seriously introduced in a 1993 review by Alan Emrich of a space oriented strategy game, *Master of Orion*:

> I give *MOO* a XXXX rating because it features the essential four X's of any good strategic conquest game: EXplore, EXpand, EXploit, and EXterminate. In other words, players must rise from humble beginnings, finding their way around the map while building up the largest, most efficient empire possible. Naturally the other players will be trying to do the same, therefore their extermination becomes a paramount concern. A classic situation, indeed, and when the various parts are properly designed, other X's seem to follow. Words like EXcite, EXperiment and EXcuses (to one's significant others) must be added to a gamer's X-Rating list.[7]

With respect to space-related strategy games that simulate social behavior, EXterminate is a common if unpleasant feature, and we might prefer 4X to signify: EXplore, EXpand, EXploit, and EXperiment. Or: EXplore EXciting EXtraterrestrial EXotics.

5.2 Multiple Solar Systems

Remarkably, a new version of *Master of Orion* became available in 2016 on Steam, the influential computer game distribution service, along with three earlier versions. Wikipedia describes it thus: "*Master of Orion: Conquer the Stars* is a turn-based strategy game that lets players take control of one of 10 playable races who can compete or coexist with other AI controlled opponents across vast galaxies. In the game, players manage their empire, colonies, technological developments, ship design, inter-species diplomacy, and combat. In *Master of Orion*, players typically begin the game with a colonized home world in their solar system. Utilizing colony resources such as Credit Income, Research, Food, and Production, the player can grow their empire, expand their fleet, explore new worlds, increase their resources and trade, and combat other races. Victory is achieved by eliminating all other opponents, winning a vote for peaceful unification or having the highest score at the end of the last turn."[8]

While winning in an open competition is the goal, much of the work to achieve it models colonization of planets, interstellar travel, and communication with extraterrestrial intelligences. Steam advertised *Master of Orion* in these scientifically relevant terms: "Forge an empire in a universe where population growth is stripping away planetary resources. Colonize unknown planets and trade with other races for their

[7] Alan Emrich, "MicroProse's Strategic Space Opera is Rated XXXX," *Computer Gaming World*, 1993 (September, issue 110): 92–93, p. 92.

[8] en.wikipedia.org/wiki/Master_of_Orion:_Conquer_the_Stars, accessed May 2017.

Table 5.1 Ten cultures competing for interstellar dominance

Species	Original description	Most recent description	Trial 1	Trial 2
Alkari	Superior pilots	Lofty, inflexible, honorable		X
Bulrathi	Terrific ground fighters	Headstrong, territorial, ferocious		X
Darlok	Supreme spies	Stealthy, treacherous, scavenging		X
Human	Expert traders and magnificent diplomats	Diplomatic, stubborn, charismatic	X	
Klackon	Increased worker production	Tireless, uncreative, hive-minded	X	
Meklar	Enhanced factor controls	Industrious, erratic, unpredictable		X
Mrrshan	Superior gunners	Fearless, warlike, proud		X
Psilon	Superior research technologies	Brilliant, unsympathetic, creative	MAIN	MAIN
Sakkra	Increased population growth	Numerous, brutish, inarticulate	X	
Silicoid	Immune to hostile planet environments	Resistant, xenophobic, withdrawn		X

knowledge." Therefore, with no hope of "winning," I twice explored the solo-player version from the vantage of what appeared to be the most intellectual of the ten main species, the Psilon, changing only the set of other species inhabiting the galaxy. Table 5.1 lists them, with descriptions from the original and most recent versions of *Master of Orion*.

At the beginning, a solo player is given many choices, including selection of the main species through which the user plays and how many from 1 to 6 of the other species will be operated by the computer, following complex, hidden algorithms. The current version shows detailed if somewhat cartoonish pictures of the species, and the Psilons are small, lean humanoids with large heads, big eyes, gray skin, and four arms. Depicting them as small and lean symbolizes their non-violent nature. But it also connects to a characteristic represented by a variable in one of the algorithms, that as natives of a low-gravity planet they have trouble colonizing planets with higher gravity. Their physical inferiority is reflected in a security variable the game says is 20% below the average, and their mental superiority by high scores on variables called *technologist* and *creative*, giving them a 50% advantage in researching new technology. The gray skin may reflect their unemotional character. The large head and big eyes express intellectuality and curiosity. The four arms may relate to their engineering skills. I wonder if the name Psilons was suggested by the robot Cylons in the 1978 television series *Battlestar Galactica*, adding a hint of psychology to stress the artificial quality of their natural cognition. The current version of *Master of Orion* explains further:

> Under the supervision of the Controller at Mentar, a sprawling network of Psilon communities comprise the Quanta. Devoted to the unraveling of the mysteries of the universe, each Quantum investigates independently in order to contribute to the knowledge of the whole. The Psilon are a peaceful race, and only keep fleets of highly advanced warships as defensive measures. Brilliant researchers, they are often ahead in the technological race, yet seldom seem to take advantage of their superiority. Instead, they prefer to trade for new discoveries while shielding themselves from the rest of the galaxy in order to continue investigating in peace.

In addition to selecting the species for the particular simulation of their social behavior, the interface offers several other starting decisions. Should the shape of the galaxy be circle, spiral, or cluster? Should the galaxy be small, medium, large, or huge? Should it be young, of average age, or old? A fourth astronomical setting concerns the planet density per star, from a low average number of planets in each solar system to a high number. Another variable, "starting age," determines how advanced interstellar flight technology would be at the beginning. The difficulty and pace of the simulation can also be set, on two six-choice scales. Also, a number can be entered in a field called "big bang seed," presumably setting the random number generator that will update itself according to a hidden function on each application. When the first trial began, this field had the number 52755 in it, so I manually entered that same number for the second trial. The software resets the seed every time the "new game" option is selected from the main menu, and usually random number generators take their seeds from the clock on the user's computer.

A thorough study of a complex system like *Master of Orion* would require using it many times, following a carefully developed experimental design, perhaps combined with observations of regular players and interviewing the system's creators. Here we do a much more modest reconnaissance. The action consists of a series of turns taken by the user and by each of the other species, whether they are operated by the software as in our two trials, or by other people. On each turn, the user may initiate several kinds of action, but mostly either moving spaceships and performing actions with them, adjusting conditions on each planet especially to produce various technological artifacts, or a combination of both. In neither of the trials was I a winner, but of course the goal was to observe the range of concepts coded into the software rather than vanquish anyone. I voluntarily ended Trial 1 with 16 colonized planets at turn 500, while Trial 2 ended on turn 436 with the total extermination of the Psilon species.

Trial 1 began when the Psilons had established a modest colony on the planet Mentar II, and had just launched a small fleet of spaceships, consisting of a colony ship carrying colonists, a frigate to defend them, and two small scout ships. The display showed this as a map of the Mentar solar system, with its central star and an asteroid field, as well as Mentar II. Later, it was possible to assemble the asteroids into a planet called Mentar Prime. But this could be done only after the colony on Mentar II had produced the necessary equipment, called a *space factory*. By turn 7, the fleet had reached an adjacent solar system, called Ezore, which already had three planets: Ezore Prime, Ezore II and Ezore III. In this nomenclature, the planet with the smallest orbit is called Prime, and the others are numbered outward. Thus, when

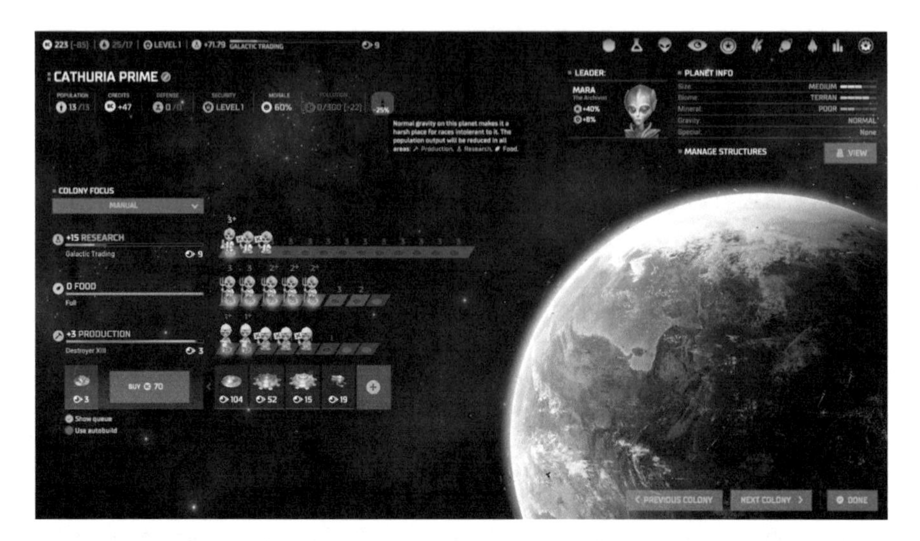

Fig. 5.1 The information display for the planet Cathuria Prime

the asteroids in the Mentar system were assembled into a planet, it would be Mentar Prime, because its orbit is inside that of Mentar II. The plot thickened at this point, because it turned out that one of the Ezore planets was already inhabited, and in the opposite direction space pirates were headed toward the Psilon colony on Mentar II. The Psilons successfully defended themselves, were able to colonize the planet Cathuria Prime in a system the opposite direction from Ezore, and were building defensive capabilities in both the Cathuria and Mentar systems.

Figure 5.1 shows the data display around turn 360 of Trial 1, for the planet Cathuria Prime, that then was the most populated Psilon world, having 13 population units. An image of the planet is in the lower right, and the rest of the display presents information about its society and economy. As printed here, many of the details may not be visible, but the standard format of the display provides a good basis for verbal description. The large section of the display on the left, in the middle and bottom, represents as horizontal bars the division of labor of the population across three functions: research, food, and production. Part of each bar includes tiny pictures of Psilon people, 3 for research and 5 each for food and production. One might infer that the population of the planet is only 13 individuals, but probably these are population units, plausibly totaling 13,000,000. Two of the research Psilons, and 3 of those in production, are sitting down rather than standing, brandishing signs with red Xs on them. These 3/13 of the population are on strike, because as one of the numbers near the top left reports, the morale of the population is only 60%. This fraction of the population has become outraged at the high taxes imposed by the government, a reflection of the fact that this social simulation includes variables representing the condition of the population, which can also be affected by problems like pollution and excessive gravity of a colonized planet.

The lowest band of images in the lower left of Fig. 6.1 is the list of things to be produced, at the far left a tiny image of the destroyer type of military spaceship, which has 3 more turns to go before it will be produced, unless the user is willing to spend 70 units of valuable tax money to produce it immediately. When the destroyer is finished, a more powerful cruiser will go into production, followed by a colony ship, a civil transport, and a space factory. Civil transports deliver one population unit of Psilons to whatever already colonized planet it is sent to, population that were not taken from another planet but somehow produced by implicit biological reproduction. Information about the planet itself is given in the upper right corner. Cathuria Prime is of medium size, with a Terran biome, poor mineral resources, and normal gravity.

By the time Trial 1 had reached turn 500, all planets had been discovered and all convertible asteroid fields had been assembled into planets, by one or another of the four intelligent species. One sector of the galaxy had been hidden to the Psilons, because they did not have good relations with the Humans who occupied it. So at moderate cost of effort and resources, the Psilons entered into diplomatic communications with the Humans, leading to a treaty that allowed the Psilons to access information about all the solar systems. Altogether, there were 40 solar systems, plus two wormholes at opposite ends of the galaxy, in the form of solar systems but facilitating long distance travel. Nominally there were 95 planets, although 6 of them were asteroids in different solar systems that could not be converted into planets, because they were connected to a story about a mysterious Antaran civilization. Instead, I had set up an asteroid laboratory at one of them, and reportedly owning 4 of these would open a new portal.[9]

It was relatively easy to copy data about the 89 planets into a spreadsheet and examine the distributions of their characteristics. With respect to size, 13 were tiny, 14 small, 29 medium, 20 large, 11 huge, and 2 were giant. Surface gravity had three categories: 21 low gravity planets, 53 normal, and 15 high. The two variables correlated with each other, but not perfectly. Both giant planets, 6 huge, 4 large, and 3 medium planets had high gravity. The 21 low gravity planets were a combination of medium, small. and tiny. However the 53 normal gravity planets included all sizes except giant: 5 huge planets, 16 large, 21 medium, 5 small, and even 6 tiny planets. Of course, in the real universe we would not expect a perfect positive correlation between diameter and surface gravity, given that the density of the planet is also a key factor. For example, the surface gravity of low-density Saturn is almost exactly the same as that of the much smaller high-density planet Earth.

However, the complex relationship between size and gravity in *Master of Orion* reflects the complex contingencies of a game that seeks to offer many complex combinations of conditions. When Psion, who are from a relatively low-gravity planet, colonize a relatively high density planet they have trouble working and their health or morale may be harmed. Therefore they needed to devote several turns of their production system to construction of gravity generators that counteract the bad effect of the planet's natural gravity on the population, research, and production of

[9]steamcommunity.com/app/298050/discussions/0/154645427523126527, accessed May 2017.

equipment or food. Conversely, giving a diversity of planets normal gravity allowed them to be exploited and comfortably experienced without the chore of building gravity generators. This is a rather precise example of a very general principle we see throughout this book: Some features of simulations may be astronomically unrealistic because they enable human users to concentrate on other features of the simulations, especially those having to do with conventional human life.

Two other somewhat independent variables, minerals and biome, described the environments of the planets. A five-category scale measured how good the mineral deposits on a planet were: 8 planets were ultra-poor, 15 poor, 42 abundant, 15 rich, and 9 ultra rich. With respect to biome, two gas giant planets were the least hospitable, and a simple graph assigned them to the zero level of biome quality. Level 1 was barely survivable, including 28 barren planets, 13 radiated, and 3 volcanic. Construction of a planetary radiation shield reduces cosmic and solar bombardment, rendering a planet barren, which seems slightly better but within the same dismal category. Biome level 2 contained 4 desert planets and 8 tundra planets. Level 3 consisted of 6 arid planets and 3 swamp planets. The very favorable level 4 planets included 1 that was tropical, 3 cavernous, 4 ocean, and 8 terran worlds like our own Earth. The perfect biome, found on only 6 worlds, was called *Gaia*, obviously in honor of the Gaia Hypothesis about nature's tendency to evolve toward perfection. As the first chapter of this book reported, James Lovelock, co-originator of the Gaia Hypothesis, was associated with the abortive Argo Venture simulation of planetary colonization.

Much of the time, the user interface is filled with a map of the section of the galaxy where action involving the user's species is taking place. Figure 5.2 is a somewhat peaceful vision of one end of the Psilon territory, on turn 406 of Trial 1. The Cathuria solar system is outside the view, to the left, and we see just a tenth of the 40 systems. To make the image more intelligible, I replaced the faint names of the four systems with bold, white letters, and highlighted flight paths between systems, with bold white lines. At this point in history, the Psilons had colonized the Mentar, Ezore and Lengzin systems, and a second plague of pirates periodically invaded from Celtsig. The only way to enter or exit the Ezore system is by way of Mentar, and either flight path from Mentar leading outside the picture goes to Cathuria, but only after passing through another solar system.

The structure of the galaxy is rather like a sociogram in sociology, in which direct communication is possible only between nodes of the network that are connected. In fact, there are just 54 flight paths linking the 40 solar systems, generally rather short, replicating the structure of interstellar travel assumed in Chap. 3 of this book, a network of relatively short journeys. However, many of the nearby systems are not connected, notably no connections between Ezore and either Lengzin or Celtsig. Given that mathematically there are 780 possible links among 40 nodes, the network density is only about 5%. For purposes of enhancing the competition between civilizations simulated by *Master of Orion*, the low density of the network connecting solar systems complicates aggressive strategies.

The two flight paths leading from Mentar outside the image, and the two leading down from Lengzin, are the only routes enemies could take to attack these Psilon worlds, through which an enemy would need to pass to attack Ezore. Each path

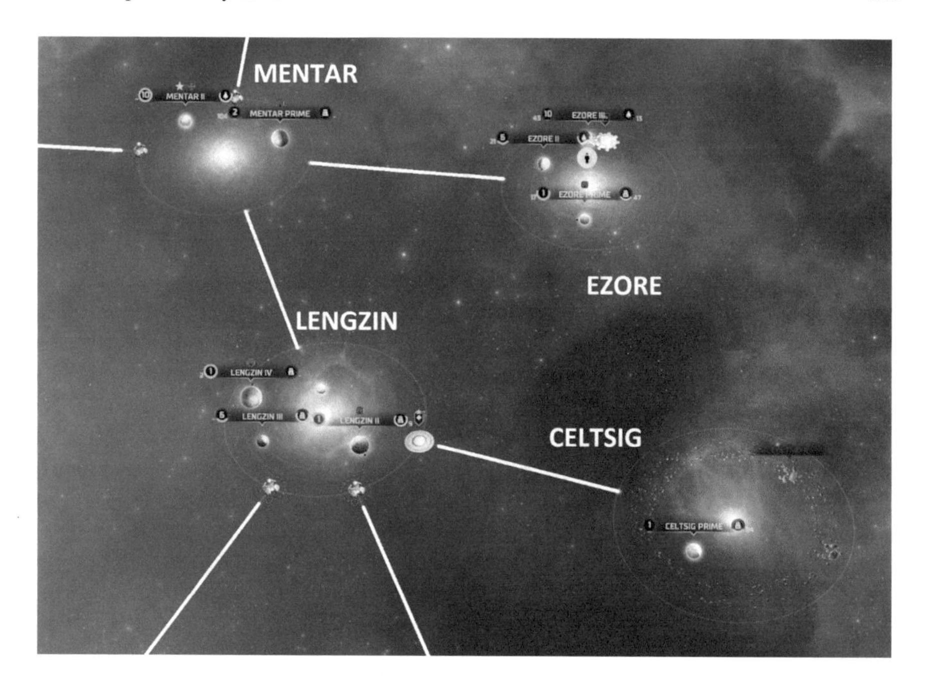

Fig. 5.2 Four connected solar systems in the Orion galaxy

connecting leads from an outer orbit containing no planet, where it is possible to build a military outpost, and indeed the Psilons have constructed one at each of these four points. None was built on the flight paths between their three system. In the outer orbit of Lengzin, on the path to Celtzig, they placed a fleet of military spaceships, strong enough to vanquish the occasional pirate invaders. On later turns, a Psilon fleet vanquished the pirates in Celtsig, and colonized its planets.

Trial 1 was distinguished by much diplomatic negotiation among the four civilizations, and occasional election votes eventually deciding that one civilization would rule the others, hopefully in peace. The elections were between Klackon and Human candidates, because they had the richest and most powerful empires. Eventually the Klackon won. The Psilon were able to progress somewhat under Klackon domination, but perfect peace was not achieved, and by turn 500 the Sakkra were extinct. The situation by then was certainly complex, but the over-all point structure was painfully clear. The Klackon had 68,251 points, the Humans had 21,888 and the Psilons had only 14,451. This was not a fair test of the cultures of the four civilizations, because other factors intervened, but with some effort, *Master of Orion* can be used as a laboratory in which to simulate the interaction of competing social theories.

Trial 2 had a very different flavor from that of Trial 1, and it never got to the point of elections. Diplomatic relations were tenuous at best. This may have been the natural result of a galaxy with 7 civilizations rather than only 4. The Psilons started out in a rather similar situation, but had great difficulty building infrastructure or colonizing planets, constantly under attack. The final turns were quite pathetic. The Psilon controlled only two solar systems, Mentar and Lengzin. Having few defenses, they knew they could not construct warships fast enough to win any battle, so every capable planet began building colonization ships, hoping that at least one of them could sneak past the enemies and find safe harbor elsewhere in the galaxy.

The Psilons made peace with the Darlok, but the Alkari declared war. Their leader, Redwing, proclaimed: "We Alkari are to rule over every horizon, even yours. It is the will of the Sky God." Two Psilon colony ships were about to escape from Lengzin, when an Alkari fleet of 30 ships surrounded them: 1 battleship of rating IV, 2 XIII-rated cruisers, 7 XVI destroyers, 2 XV destroyers, 10 XV frigates, 3 XIV frigates, 4 XIII frigates, and 1 troop transport of rating V. The colonists vanished without a trace, and soon the Alkari had destroyed the last Psilon colony.

The galactic war had not finished, but lacking a "living" avatar, it was impossible to learn the ultimate outcome. In the last available statistics, two species were nearly tied, the Silicoid with 28,719 points, and the Alkari, with 28,207. The Meklar had 22,587, the Darlok 18,080, and the Bulrathi 11,026. True scholars that they were, the Psilon might have wondered if the Bulrathi took their name from the Kilrathi, the prominent extraterrestrial species from the classic *Wing Commander* game dating from 1990.[10] But no Psilon scholars still existed, having become extinct along with the Mrrshan.

5.3 Cultural Relativism

Remarkably, many of the strategy games about galactic colonization imagine a collection of competing civilizations, each of which possesses a culture based on a coherent set of principles. In a very real sense, this is a reflection of the perplexing first book by Proto-existentialist philosopher, Friedrich Nietzsche, *The Birth of Tragedy*. There he contrasted three culture types that were equivalent to personality types: Apollonian, Dionysian, and Buddhist. The *Apollonian* is cool, rational, classical, and when it does not speak in grammatical sentences expresses itself through the visual arts. The *Dionysian* is hot, lustful, romantic, and when it does not roar with animal noises expresses itself through music and dance. Apollonianism emphasizes the individual while Dionysianism emphasizes the group. Buddhism de-emphasizes both the individual and group in meditative withdrawal from the world. Later, in *Thus Spake Zarathustra*, Nietzsche attempted to transcend these three cultural archetypes, and we can imagine a fourth culture-personality pattern that emphasizes both the individual and the group, calling it *Zarathustran*, although the tragic philosopher failed to attain it (Nietzsche 1872).

The most influential cultural typology found in a strategy game simulating interstellar social relations is the triad in *StarCraft*: Protoss, Zerg, and Terran (Human), which I briefly explored in an earlier book (Bainbridge 2011). The Protoss are a

[10]www.wcnews.com/wcpedia/Kilrathi, accessed May 2017.

highly-advanced aloof species that developed a strict quasi-religious philosophical discipline to control its own internal violence, unsympathetic to species that lack that creed, reminiscent of Nietzsche's Apollonian type of culture. In contrast, the Zerg are a quasi-intelligent hive species that swarms into new territory, overwhelming and consuming the natives, thus Dionysian. The humans are an aggressive and naturally duplicitous species, in which leaders urge everyone to work for the common good while exploiting their followers for personal gain. Wikipedia summarizes the backstory:

> Millennia before any of the events of the games, a species known as the Xel'Naga genetically engineer the Protoss and later the Zerg in attempts to create pure beings. These experiments backfire and the Xel'Naga are largely destroyed by the Zerg. Centuries before the beginning of *StarCraft* in 2499, the hardline international government of Earth, the United Earth Directorate (UED), commissions a colonization program as part of a solution to overpopulation. However, the computers automating the colony ships malfunction, propelling the Terran colonists far off course to the edge of Protoss space. Out of contact with Earth, they form various factions to maintain their interests. Intrigued by the behavior and mentality of the Terrans, the Protoss remain hidden to examine the humans, while protecting them from other threats without their knowledge. However, the Zerg target the Terrans for assimilation to harness their psionic potential, forcing the Protoss to destroy tainted Terran colonies to contain the Zerg infestation.[11]

In gamer lingo, *zerg* has become the common term for a mass blitzkrieg attack, in which swarms of low-level military assets are rapidly sacrificed in an attempt to destroy the enemy before it could organize an effective defense.[12] *StarCraft* with its three civilizations was released in 1998, and in 1999 an historically significant interstellar game was released that simulated seven civilizations, *Sid Meier's Alpha Centauri*. It was an offshoot of the *Civilization* series that launched back in 1991, and has remained true to the XXXX tradition (Ford 2016). Later we will closely examine its recent manifestation, the 2014 strategy game, *Sid Meier's Civilization: Beyond Earth*.

Rather commonly, strategy games are based loosely on real terrestrial history, for example the *Total War* series, that includes *Rome Total War* which simulates many real battles in the history of ancient Rome. Sometimes, a single individual historical figure plays a key role, such as Napoleon Bonaparte in *Napoleon: Total War*. In the case of *Sid Meier's Alpha Centauri*, the player selects which of 7 fictional social movements of the future to join, none exactly resembling a particular historical example, yet terrestrial in origin and having a Napoleonic leader. They are listed in Table 5.2 with descriptive terms, the leader's name and nation of origin, and a quotation for the leader found inside the game.

Each movement is represented by a charismatic leader, indeed is an extension of that individual's personality (Weber 1978). Conversely, each may be seen as a philosophical archetype, rendered intelligible to the average player by being personified as the charismatic leader (Bainbridge 2017). One source of information about each is a text file inside the folder containing the game software, that the game accesses

[11] en.wikipedia.org/wiki/StarCraft, accessed May 2017.

[12] en.wikipedia.org/wiki/Rush_(video_gaming), accessed May 2017.

Table 5.2 Seven social movements with interstellar ambitions

Movement	Leader	Quotation ("Blurb")
Gaia's Stepdaughters (ecological harmony)	Lady Deirdre Skye (Scotland)	In the great commons at Gaia's Landing we have a tall and particularly beautiful stand of white pine, planted at the time of the first colonies. It represents our promise to the people, and to Planet itself, never to repeat the tragedy of Earth
Human Hive (totalitarian legalism)	Chairman Sheng-Ji Yang (Great China)	Learn to overcome the crass demands of flesh and bone, for they warp the matrix through which we perceive the world. Extend your awareness outwards, beyond the self of body, to embrace the self of group and the self of humanity
Lord's Believers (dominionism)	Sister Miriam Godwinson (Christian States of America)	The righteous need not cower before the drumbeat of human progress. Though the song of yesterday fades into the challenge of tomorrow, God still watches and judges us. Evil lurks in the datalinks as it lurked in the streets of yesteryear
Morgan Industries (economic, capitalist)	CEO Nwabudike Morgan (Namibia)	Human behavior is economic behavior. The particulars may vary, but competition for limited resources remains a constant. Need as well as greed have followed us to the stars, and the rewards of wealth still await those wise enough to recognize this deep thrumming of our common pulse
Peacekeeping Forces (democracy)	Commissioner Pravin Lal (India)	Free flow of information is the only safeguard against tyranny. The once-chained people whose leaders at last lose their grip on information flow will soon burst with freedom and vitality, but the free nation gradually constricting its grip on public discourse has begun its rapid slide into despotism
Spartan Federation (militaristic)	Colonel Corazón Santiago (Puerto Rico)	Superior training and superior weaponry have, when taken together, a geometric effect on overall military strength. Well-trained, well-equipped troops can stand up to many more times their lesser brethren than linear arithmetic would seem to indicate
University of Planet (technocracy)	Academician Prokhor Zakharov (Russia)	The substructure of the universe regresses infinitely towards smaller and smaller components. Behind atoms we find electrons, and behind electrons quarks. Each layer unraveled reveals new secrets, but also new mysteries

for each of the seven. Other sources include various online wikis, which occasionally link to Wikipedia to connect the game to real-world scholarship.[13] For example, Lord's Believers follow a religious approach to governance: "Dominion Theology (also known as Dominionism) is a group of Christian political ideologies that seek to institute a nation governed by Christians based on Christian understandings of biblical law."[14] Or to cite a contrasting example, Human Hive advocates Chinese legalism: "Fǎ-Jiā or Legalism is one of the six classical schools of thought in Chinese philosophy that developed during the Warring States period. Grouping thinkers with an overriding concern for political reform, the Fa-Jia were crucial in laying the 'intellectual and ideological foundations of the traditional Chinese bureaucratic empire', remaining highly influential in administration, policy and legal practice in China today."[15]

A more recent example, in many ways comparable to *Master of Orion*, is *Stellaris*, dating from 2016 and set in 2200. The online service, Steam, advertises it thus: "Your species has mastered the seemingly impossible. Faster than light travel means a new era for your civilization. Brave pioneers set forth from our ancient homeworld into the unknown, while scientists unlock more and more of the vast mysteries of the cosmos… But you are neither the first, nor the only, species to climb to the heavens. You must test your military and diplomatic prowess against rival galactic empires." *Stellaris* features 10 factions, any of which may be selected by the player, two of them human and the other eight, extraterrestrial. As found in the game's faction-selection screen, Table 5.3 lists their names and main characteristics.

The innate characters of the species are reflected in relevant algorithms. For example, the reptilian Tyznn and birdlike Yondar both are "rapid breeders," which means that less time is required for their population to increase. In contrast, the Jehetma are slow breeders, and the other seven are not remarkable in terms of their rate of reproduction. Both of the terrestrial "humanoid" groups are wasteful, which suggests that sometimes they require more resources to accomplish a goal. Other characteristics are not innate, but were defined during the historical process by which the particular social organization arose.

The two terrestrial groups do not differ at all in their biology or innate character, but do differ in their social organization and values. The United Nations of Earth consider themselves a beacon of liberty based on an idealistic foundation, both xenophile and fanatically egalitarian. The Commonwealth of Man possesses nationalistic zeal led by a distinguished admiralty, both xenophobic and fanatically militarist. The *Stellaris* wiki explains the distinction between xenophile and xenophobe orientations: "A species may or may not equate with an empire. At first every empire will be made up entirely of its original species but eventually it may include other species through

---

[13] civilization.wikia.com/wiki/Factions_(SMAC), strategywiki.org/wiki/Sid_Meier%27s_Alpha_Centauri/Factions, accessed May 2017.

[14] en.wikipedia.org/wiki/Dominion_Theology, accessed May 2017.

[15] en.wikipedia.org/wiki/Legalism_(Chinese_philosophy), accessed May 2017.

Table 5.3 A diversity of interstellar competitors

Name	Government	Biology	Innate character
United Nations of Earth	Representative Democracy	Humanoid	Adaptive, nomadic, wasteful
Commonwealth of Man	Constitutional Dictatorship	Humanoid	Adaptive, nomadic, wasteful
Tzynn Empire	Star Empire	Reptilian	Strong, talented, rapid breeders, nonadaptive
Kingdom of Yondarim	Divine Empire	Avian	Sonformists, talented, solitary
Ix'ldar Star Collective	Totalitarian Regime	Arthropoid	Rapid breeders, communal, repugnant, strong
Chinorr Stellar Union	Science Directorate	Molluscoid	Industrious, repugnant, natural engineers
Jehetma Dominion	Irenic Bureaucracy	Fungoid	Slow breeders, thrifty, communal
Scyldari Confederacy	Moral Democracy	Mammalian	Charismatic, solitary, natural sociologists, quick learners
Kel-Azaan Republic	Citizen Republic	Arthropoid	Adaptive, nomadic, fleeting
Blorg Commonality	Military Junta	Fungoid	Repugnant, solitary, venerable

conquest or inter-empire migration. Xenophile empires particularly favor this idea while xenophobe empires hate it."[16]

5.4 An Advanced Civilization

In 2014, the successor to *Sid Meier's Alpha Centauri*, titled *Sid Meier's Civilization: Beyond Earth* because it was also the latest entry in his *Civilization* series:

> All titles in the series share similar gameplay, centered on building a civilization on a macro-scale from prehistory up to the near future. Each turn allows the player to move his or her units on the map, build or improve new cities and units, and initiate negotiations with the computer-controlled players. In between turns, computer players can do the same. The player will also choose technologies to research. These reflect the cultural, intellectual, and technical sophistication of the civilization, and usually allow the player to build new units or to improve their cities with new structures. In most games in the series, one may win by military conquest, achieving a certain level of culture, building an interstellar space ship, or achieving the highest score, among other means.[17]

[16]www.stellariswiki.com/Species, accessed May 2017.

[17]en.wikipedia.org/wiki/Civilization_(series) , accessed December 2017.

This section will concentrate on the moment-to-moment social interactions surrounding *Beyond Earth*, but some introduction about the social structures built into it and the experience of entering briefly are needed as introduction. The version I explored included the first main expansion, *Rising Tide*, which added four civilizations to the original eight to which a player could belong, and changed some of the statistical measures fundamental to the simulation of social behavior. Table 5.4 lists the current dozen civilizations, identifies which region of our planet each represents, gives the name of the players' first city which serves as a capital, names the leader of the particular group, and gives two characteristics of the leader or civilization that were parameters of the action algorithms.

The four civilizations added in the expansion were Al Falah, Chungsu, INTEGR, and the North Sea Alliance. An interesting detail about Al Falah is that its leader is a woman, despite the fact that many of today's nations in the Middle East are unusually male-dominated. Chungsu, the new Korean group, differs from the existing Pan-Asian Cooperative in that it is ocean-oriented and can build its first city in the sea of the new planet. INTEGR is an abbreviation for Initiative für Nachhaltige Technologien, Effizienz, Gerechtigkeit und Rechtschaffenheit (Initiative for Sustainable Technologies, Efficiency, Justice, and Righteousness).[18] Like Chungsu, the North Sea Alliance can build its first city in the sea, and derives from "the islands of Britain, Ireland and the present-day Nordic region, which includes the island of Iceland and the peninsulae of Scandinavia and Jutland and the Netherlands and the small islands in the proximity of the two peninsulae."[19]

The column of Table 5.4 labeled *personality* is a characteristic of the leader that functions as a supreme value of the group's culture. For example, the first one, Asabiyyah, is an Arab concept of social solidarity "with an emphasis on unity, group consciousness and sense of shared purpose, and social cohesion."[20] The result within the simulation is that development of Al Falah's cities has a yield 150% of the typical civilization, and this factor can be increased.[21]

The column labeled Affinity reports the original configuration, prior to Rising Tide, because the expansion separated this variable from the civilization, and allowed the player to decide which to invest in, even combining pairs of the three. Here is how they were described on an information page at the beginning of my brief exploration:

[18]civilization.wikia.com/wiki/INTEGR_(CivBE), accessed June 2017.

[19]civilization.wikia.com/wiki/North_Sea_Alliance_(CivBE), accessed June 2017.

[20]en.wikipedia.org/wiki/Asabiyyah, accessed June 2017.

[21]civilization.wikia.com/wiki/Personality_traits_(Rising_Tide), accessed June 2017.

Table 5.4 A dozen terrestrial civilizations with extraterrestrial strategies

Name	Constituency	Capital	Leader	Personality	Affinity
Al Falah	Middle East	Ard	Arshia Kishk	Asabiyyah	
American Reclamation Corporation	North and Central America	Central	Marjorie Fielding	Corporate Espionage	Purity
Brasilia	South America	Citadela	Rejinaldo Bolivar De Alencar-Araripe	Guerrilla Mastery	Purity
Chungsu	Korea	Jeongsang	Han Jae-Moon	Jeog-ui Naebu	
Franco-Iberia	France, Spain, Portugal, Italy, Northern Africa	Le Coeur	Élodie	Shining Path	Unknown
INTEGR	Germany	Weltgeist	Lena Ebner	Kategorischer Imperativ	
Kavithan Protectorate	India, Pakistan, South Asia	Mandira	Kavitha Thakur	Doctrine of Oneness	Harmony
North Sea Alliance	Nordic countries, Great Britain	Deepcastle	Duncan Hughes	Indefatigable	
Pan-Asian Cooperative	China, Korea, Japan, Vietnam	Tiangong	Daoming Sochua	Thousand Hands	Supremacy
People's African Union	Sub-Saharan African States	Magan	Samatar Jama Barre	Umoja	Harmony
Polystralia	Australasia, Polynesia, Maritime SE Asia	Freeland	Hutama	Common Bond	Unknown
Slavic Federation	Russian Federation, Eastern Europe, Balkans	Khrabrost	General Kozlov	Cosmonaut Legacy	Supremacy

Purity: Reveres and preserves our origins on Old Earth, and reshapes the new planet in its image - as ideal home for an ideal human race.

Supremacy: Looks neither to the old or new world, but draws on human ingenuity and technology to transcend to a new, independent existence.

Harmony: Embraces and learns from the new planet, merging the lessons into itself to evolve into a new form of humankind.

For the following illustration, I selected the Franco-Iberian civilization, and in the early stages of colonization was fascinated to find this historical note: "Élodie is considered one of, if not the most important art critic and historian of the past 300 years. Her writings on the video game era of the early 21st century are considered the most authoritative take on the subject, and her discussion on the death of the internet sparked an entire artistic movement. Her greatest critical work is considered by many to be the marriage of the two topics in her Critique of Isotopic Decay 3."

A main algorithmic effect of Élodie is to increase the morale of her simulated followers by 10%. Her Shining Path personality allows the player to acquire *virtues* at a cheaper cost in *culture* points. Culture "represents the power of social innovation and influence" and is earned through such actions as constructing buildings, expeditions, and setting up trade routes.[22] As a general *Civilization* wiki explains, "Virtues are an orthogonal system to complement and augment your technology choices and play style. Whereas a player's technology path may be a series of adaptive choices to the environment and circumstances, Virtues represent the player's long term choices about style and flavor which are more removed from immediate circumstances."[23] By orthogonal, the wiki means that the player can decide which virtue tree to invest in, so the Shining Path permits quicker attainment of a goal, without telling the player what that goal should be. The four virtues are Might, Prosperity, Knowledge, and Industry.

Another choice at the beginning of *Beyond Earth* is the kind of planet to be colonized. As well as the option to design a planet myself, I was given three pre-designed choices: (1) Kaku-159 e is a "Fungal Terran World: A world with a few large landmasses separated by oceans and some smaller islands." (2) Jones-178 d is a "Lush Protean World: A world of one ocean and one very large, continuous landmass with the possibility of small, coastal islands." (3) Springer-569 g is a "Lush Atlantean World: A world of islands of varying sizes separated by narrow water passages." Another time the gross choices would be the same, but the names would change, so that one can have a sense of exploring ever more planets. I chose the lush protean choice, landed on a seacoast of the one continent, and set up the initial outpost that would grow into the city of Le Coeur - "the heart" - as shown in Fig. 5.3.

This high resolution screenshot shows Le Coeur, still tiny at this point, in the lower left corner of the picture, defended by a contingent of soldiers. Over to the right of center, a hexagon highlights an expedition I had sent out to begin exploring, and two sets of blurry objects to the right of it are colonies of giant scarab beetles. This particular planet did not have a native intelligent civilization, but it did have monstrous lifeforms. Soon, expeditions sent by the other eleven Earth factions would begin landing and setting up competing colonies.

The map of each world in *Beyond Earth* is composed of hexagons, reminiscent of the squares in jetan, across which the player may move units that represent teams of colonists, soldiers, and scientists. Hexagons differ in their natural resources and also in environmental hazards, such as the health-destroying miasma depicted as a

[22]civilization.wikia.com/wiki/Culture_(CivBE), accessed June 2017.

[23]civilization.wikia.com/wiki/Virtues_(CivBE), accessed June 2017.

Fig. 5.3 A very early state of planetary colonization

greenish haze on many hexagrams in this picture. The interface hides the hexagon borders, except in areas where action decisions may be taken, and in distant areas that have not yet been explored. For example, we do not know whether the hexagons in the upper right contain land or water, and that will become clear only when one of the player's units approaches.

Figure 5.4 shows two scenes soon after the screenshot in Fig. 5.3, and separated by only two turns. The upper image shows the soldiers, highlighted in a hexagon showing that action is being taken, shooting at giant scarab beetles that are attacking from the right. Immediately below them the expedition waits. Off on the peninsula to the far right, a horde of wolf beetles waits for its opportunity to enter the fight. In the lower image, the expedition has moved to the now-safe hexagon previously occupied by the soldiers. The soldiers and the wolf beetles have both advanced toward each other, and are now in battle.

Beyond Earth is a simulation of large-scale social behavior, operated by a single person, but extensive social interaction surrounds it, for example on YouTube. Many players posted videos of particularly exciting moments they experienced, but a few people who seem dedicated to being YouTube stars posted very ambitious videos. Wearing a spacesuit costume, Angry Joe introduced a rather critical review, complaining the *Beyond Earth* was just a bland remake of *Civilization 5*, garnering 1,800,565 views by July 15, 2017.[24] A pair of video producers calling themselves 2BCProductions2BC and claiming to have uploaded over 3,000 videos, generated 23 videos of *Beyond Earth* playthroughs, consisting of going through the game while offering thoughtful narration explaining what is happening.[25] Watching all of them

[24] www.youtube.com/watch?v=3dk4Reefp_8, accessed June 2017.

[25] www.youtube.com/watch?v=N80sMZwm5CY&t=28s, accessed June 2017.

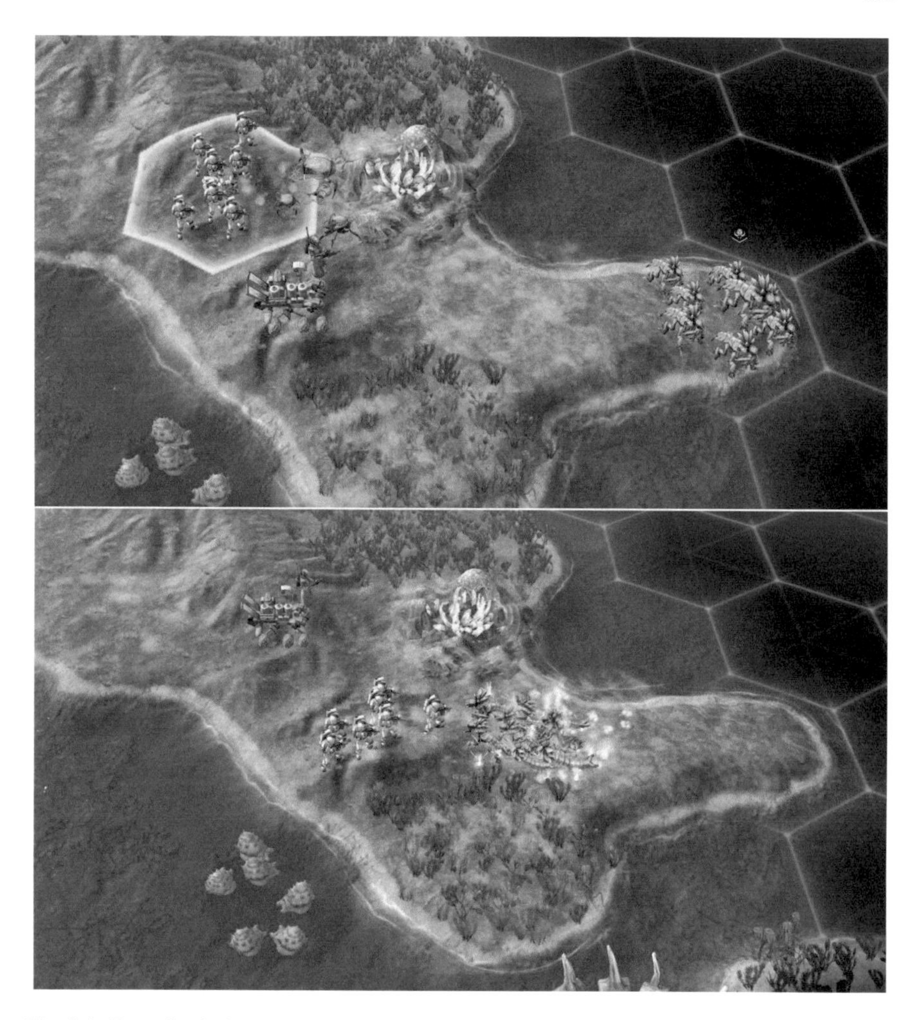

Fig. 5.4 Engaging in battle on an Alien planet

would take 11 h, and their view numbers ranged from 146,984 down to 8,766. For literally millions of people, YouTube and its competitors have become the successor of broadcast and cable television, a computer-based mass medium produced largely by amateurs.

5.5 The Definition of Social Strategy Simulations

From jetan to *Beyond Earth*, space-oriented strategy games appear to be wargames, starting from the premise that the player must have an opponent. *Buzz Aldrin's*

Race Into Space certainly fit that definition, and it simulated the real space race in which the Soviet Union and United States had competed. However, strategic planning would be required for any major spaceflight development. For example, colonization of Mars would require a strategy for attracting the investments required to transport the colonists and their equipment, logistics for exploitation of the red planet's natural resources in some sequence of mutually supportive developments, and a staged plan for expanding from a few initial outposts to major settlements. Yet, correctly Wikipedia describes strategy games in terms of human conflict and victory:

> Strategy video games are a video game genre that focuses on skillful thinking and planning to achieve victory. It emphasizes strategic, tactical, and sometimes logistical challenges. Many games also offer economic challenges and exploration. They are generally categorized into four sub-types, depending on whether the game is turn-based or real-time, and whether the game focuses on strategy or tactics... Specifically, a player must plan a series of actions against one or more opponents, and the reduction of enemy forces is usually a goal. Victory is achieved through superior planning, and the element of chance takes a smaller role. In most strategy video games, the player is given a godlike view of the game world, and indirectly controls game units under their command. Thus, most strategy games involve elements of warfare to varying degrees, and feature a combination of tactical and strategic considerations. In addition to combat, these games often challenge the player's ability to explore, or manage an economy.[26]

As in jetan, often the number of opponents is one, although that one may be the computer itself. In the early days of massively multiplayer online games, it was not entirely clear how much the emphasis should be on strategy, versus role-playing and dramatic graphics of physical combat. Soon, the marketplace decided that games like *World of Warcraft* would be most successful with thousands of players online at once, each represented by a lone avatar rather than by an army, working through interesting story quests. Strategy was relegated to secondary functions, notably in virtual economies that produced and exchanged armor, weapons, and potions that increased the avatar's chance of victory in combat with non-player boss characters or on battlefields. Some multiplayer strategy games existed, for example letting players build and defend castles in socially disorganized Medieval environments, but usually unpopular and technically unsophisticated.

To get a sense of how the genre of multiplayer strategy games related to spaceflight development, I downloaded data about all 1,263 games listed on the MMORPG.com website, then focused on the 218 sci-fi games, and among them, the 21 categorized as massively multiplayer online real-time strategy (MMORTS) games. I checked each one on its own website as well as at MMORPG.com, finding that 4 were still under development, 6 were extinct, 2 did not relate to space, and just 9 currently were in operation. The earliest pair were blends of strategy games and role-playing games, each requiring the user to download specially programmed software:

> Shattered Galaxy (2001): "Shattered Galaxy, the world's first published massively multi-player online real-time strategy game, combines the excitement of real-time strategy with the character development of role-playing in a persistent online world."

[26]en.wikipedia.org/wiki/Strategy_video_game, accessed June 2017.

> Star Sonata 2 (2004, 2011): "Star Sonata provides a rich fabric in which the players create and define the universe."

The other 7 were *browser games*, which were cheap to produce and limited in both complexity and graphics appeal, because they worked through standard web browsers like Internet Explorer or Goggle Chrome:

> Astro Empires (2006): "Astro Empires is a space strategy browser game. The game runs on a persistent universe in real time. Develop bases, research new technologies, trade with your neighbors, produce fleets, form alliances and engage into epic battles!"
>
> Nemexia (2010): "The Universe is big, we won't lie to you. But how big will it feel depends of how many Galaxies you can conquer. You can colonize the farthest galaxies and spread your might over the Universe."
>
> Empire Universe 3 (2013): "Your interstellar army awaits. Lead them to victory as you advance your career. The stars are in your hands! Become a wealthy interplanetary trader or build diplomatic alliances in your quest to rule the galaxy."
>
> AD2460 (2014): "Become more powerful by researching new technologies, developing new ship types and securing necessary resources from your mining outposts on other planets. Fight other players for control of the best resource planets the galaxy has to offer."
>
> Astro Lords: Oort Cloud (2014): "The players' home bases are asteroids, on which crystals are mined, energy produced, factories built. All those resources will be used to produce ammunitions and spare parts for their spaceships. This base is the player's primary stronghold: other Lords can raid it, but they won't be able to capture it. Players can also capture up to five additional asteroids, which will be used to extract crystals. But beware: the player will have to remain on the alert to keep other Lords from recapturing the asteroids."
>
> Galaxy Warfare (2015): "Galaxy Warfare is a grid-based space exploration and combat game. It is simple to play, but difficult to master. Players each control a ship and are free to explore a X,Y based galaxy, with planets, resources and mysteries. Players also engage in combat, form fleets, enhance their ship and systems and fight to control the galaxy."
>
> Star Colony (2015): "Welcome to StarColony, a science fiction MMO strategy. The fate of a humans' colony on the distant planet of Terra Sperata is in your hands. Get resources for construction, exploration and defense. Plan your base so that enemies have no chance of conquering it. Attack other players' colonies - let the strongest win!"

MMORPG.com does not list *Master of Orion*, *StarCraft*, *Sid Meier's Alpha Centauri*, *Stellaris*, or *Sid Meier's Civilization*: *Beyond Earth*, presumably because they do not qualify as social role-playing games, following conventional definitions. However, those definitions may be flawed, and I would like to suggest that one space-related game also not on MMORPG.com's list is a social, role-playing strategy simulation that entirely lacks combat, namely *Kerbal Space Program*.

Kerbals are somewhat comical "small green humanoids, who have constructed a fully furnished and functional spaceport called the Kerbal Space Center (KSC) on their planet Kerbin. Despite being shown as cartoonish beings sometimes lacking common sense, Kerbals have shown themselves capable of constructing complex spacecraft parts and performing experiments to realize their scientific goals."[27] When I did observation field research at KSC in December 1972, it was called Kennedy Space Center, and I watched the last human launch to the Moon. The Kerbals call

[27] en.wikipedia.org/wiki/Kerbal_Space_Program, accessed June 2017.

Fig. 5.5 Building a spaceship in the vehicle assembly building

Kerbin's moon Mun, and they allow the player to build and launch their own expedition there. In the initial tutorial one gets lectures from a brilliant space engineer named Wernher Von Kerman (Wernher von Braun) and an experienced astronaut named Gene Kerman (Gene Cernan who was on that last lunar flight). Interacting with characters like these provides a degree of role-play quality. Figure 5.5 shows a scene early in the tutorials that instruct the player on how to design and construct spaceships for special missions.

The spacecraft is suspended near the center of the image, with a crew capsule on top to be piloted by Jebediah Kerman, flanked by two research modules, atop five fuel units to provide exactly the right thrust, stabilization fins and the engine at the bottom. The background is the Vehicle Assembly Building (which I had entered at the other KSC back in 1972), with unnamed Kerbals wandering around below. The icons at the left are one of several sets of equipment that may be used to assemble vehicles designed for different missions. So the strategy is selecting and assembling engineering components to create the launch vehicle and its payload, which will operate according to a precise plan, from the firing of the first stage to the opening of the parachute.

Kerbal Space Program is a solo game, but highly social, because players communicate outside it extensively. One context for its sociability is educational, because its depiction of the physics of spaceflight is sufficiently realistic that a version is used in schools: "Experiment. Create. Explore Space. KerbalEdu is an official school-ready standalone remix of the award winning game Kerbal Space Program. The game is available for everyone to purchase and has been enhanced with features that help integrate it into the classroom. Players assist the lovable-but-hapless Kerbals as they build rockets and carry out orbital missions. Students master real-world physics and engineering skills as their understanding of STEM concepts blast off!"[28] While designed for a niche audience, *Kerbal Space Program* is extremely well-regarded. As of July 15, 2017, 40,555 people had posted reviews on Steam, 39,313 or 97% rating it positively.

A major social dimension is the extensive online forums, notably the Challenges and Mission Ideas topic: "Take on a KSP challenge, or submit one yourself."[29] Five of these challenges had been viewed at least 100,000 times:

> The K Prize - 100% reusable spaceplane to orbit and back; posted by boolybooly on May 22, 2012, with 557,363 views.
>
> The Ultimate Jool-5 Challenge: land Kerbals on all moons and return in one big mission; posted by Ziv on November 7, 2013, with 358,609 views.
>
> Duna Permanent Outpost Mission Architecture Challenge; posted by sturmstiger on May 25, 2015; with 157,499 views.
>
> SSTO/Spaceplane/Airplane Design Contest II: Akademy Awards; posted by Wanderfound on September 14, 2014; with 118,411 views.
>
> Stock Payload Fraction Challenge: 1.0.5 Edition; posted by Red Iron Crown on April 27, 2015; with 100,839 views.

The first of these challenges stimulated 2,558 replies, in which players described their attempts, often identifying specific spacecraft design problems. Among the first comments posted back on May 22, 2012, these convey the confluence of technical and human issues: "You have no idea how much I have tried to do that. Try and try, space is huge. I still haven't had a plane or ship that can do that." "Aircraft failed to return to Kerbin, the weight distribution was such that the engine flew through the airframe and the craft disintegrated, I have another more stable plane but I need to get out in this sun, it's been cold for so long here." "Heh, looks like I need to step up my spaceplane game. The one spaceplane I've gotten into orbit ran out of fuel doing so, partially due to mis-routed fuel lines, and ditched a pair of lower-atmosphere engines, an SRB [solid rocket booster], and the landing gear (attached to the engines and SRB) on the way up. (The pod decouples and lands via parachute. I'm not that mean.)" "Landing from orbit is hard with mine D: It just tumbles until it hits the ground."

5.6 Conclusion

The YouTube videos and online forums related to any of the games described in this book could be the focus of major research on popular conceptions of space exploration and its future potential for humanity. To be sure, the people who post their views in these new social media are neither random samples of the population nor

[28]kerbaledu.com, accessed August 2017.

[29]forum.kerbalspaceprogram.com/index.php?/forum/25-challenges-mission-ideas/&sortby=views&sortdirection=desc, accessed August 2017.

political leaders. However, spaceflight has always been promoted by social movements to which relatively few people belonged, so this may be true in the future as well (Bainbridge 1976). Whether intended as educational software or not, the strategy games give players experience in complex systems that require analysis and planning. Thus, they may play a significant role preparing thousands of people for the space-related opportunities for real-world accomplishment that may arrive over the coming decades.

The examples emphasized in this chapter combine strategy and role-playing, with the consequence that each culture, whether extraterrestrial or a faction of terrestrials, tends to have a distinct personality. As I noted in an essay titled "Patterns of Extraterrestrial Culture," in the middle of the twentieth century cultural anthropologists often analyzed cultures in terms similar to those used to categorize personalities (Bainbridge 2013). To some extent, this reflected the influence over that period of psychoanalysis, but it also expressed the culture of continental European intellectuals who tended to analyze social reality in terms of categories often called *ideal types* or *archetypes*, depending upon the particular school of thought. More modestly, it was reasonable to deduce that cultures varied in their *modal personality*, even if was not universal in their populations (Benedict 1934). Today, this perspective seems less influential, yet not totally obsolete, sometimes using terms like *style* or *trope* rather than *personality*. For example, writing in *Current Anthropology*, Michael Carrithers, Louise J. Bracken and Steven Emery considered the trope of attributing distinct personalities to different biological species, a human habit that may shape our orientation toward extraterrestrial civilizations (Carrithers et al. 2011). In the same journal, Eitan Wilf had the culture and personality perspective in mind when analyzing how a team of computer scientists developed algorithms to simulate the artistic styles of jazz musicians (Wilf 2013).

This suggests that a worthwhile research project could explore the ways in which personality is programmed into non-player characters in computer games. For example, a standard set of alternatives found in popular combat games like *World of Warcraft* concerns the aggressiveness of animals: A wolf may attack a player's avatar if it comes too near, and a dragon may do so at a greater distance (Bainbridge 2010). A rabbit may ignore the avatar, and flee or quickly die if attacked by the avatar. Many other animals may never initiate an attack but are ready to fight back if they themselves are attacked. Some humanoids will fight for a while, but then try to escape if they suffer much damage, just like the princess in jetan. In this chapter we noted that followers of Élodie in *Sid Meier's Civilization: Beyond Earth* had 10% greater morale than the average of other groups, which factors into the algorithm that controls their productivity. After cataloguing the concepts and algorithms that simulate various personality types in existing games, the researcher would be ready to identify human characteristics that are not included, and seek to discover algorithms that could simulate them as well.

References

Bainbridge, William Sims. 1976. *The Spaceflight Revolution*. New York: Wiley Interscience.
Bainbridge, William Sims. 2010. *The Warcraft Civilization: Social Science in a Virtual World*. Cambridge, Massachusetts: MIT Press.
Bainbridge, William Sims. 2011. *The Virtual Future,* 9–11. London: Springer.
Bainbridge, William Sims. 2013. Patterns of Extraterrestrial Culture. In *Extraterrestrial Altruism*, ed. Douglas A. Vakoch, 275–294. Berlin: Springer.
Bainbridge, William Sims. 2016. *Star Worlds: Freedom Versus Control in Online Gameworlds*. Ann Arbor, Michigan: University of Michigan Press.
Bainbridge, William Sims. 2017. Surreal Impersonation. In *Video Games and Religion: Methods and Approaches*, ed. Vit Sisler, Kerstin Radde-Antweiler and Xenia Zeiler. Abingdon-on-Thames, UK: Routledge.
Benedict, Ruth. 1934. *Patterns of Culture*. Boston: Houghton Mifflin; Opler Marvin K. (ed.). 1959. *Culture and Mental Health*. New York, Macmillan; Kaplan, Bert (ed.). 1961. *Studying Personality Cross-Culturally*. New York: Harper and Row; Wallace, Anthony F. C. and Raymond D. Fogelson. 1961. Culture and Personality. *Biennial Review of Anthropology* 2: 42–78; LeVine, Robert A. 1963. Culture and Personality. *Biennial Review of Anthropology* 3:107–145.
Burroughs, Edgar Rice. 1920. *Thuvia, Maid of Mars*, 252, 255 p. Chicago: A. C. McClurg.
Burroughs, Edgar Rice. 1922. *Chessmen of Mars,* 373–375. Chicago: A. C. McClurg.
Carrithers, Michael, Louise J. Bracken and Steven Emery. 2011. Can a Species be a Person? A Trope and its Entanglements in the Anthropocene Era. *Current Anthropology* 52(5):661–68.
Ford, Dom. 2016. eXplore, eXpand, eXploit, eXterminate: Affective Writing of Postcolonial Historyand Education in Civilization V. *Game Stdies* 16(2):1–13.
Nietzsche, Friedrich. 1872. *Die Geburt der Tragödie*. Munich: Goldmann, Originally, *Also Sprach Zarathustra* (Stuttgart: Kroner, Originally 1885); Bainbridge, William Sims 2010. Burglarizing Nietzsche's Tomb. *Journal of Evolution and Technology* 21(1):37–54, jetpress.org/v21/bainbridge.htm, Patterns of Extraterrestrial Culture, in *Extraterrestrial Altruism*, ed. Douglas A. Vakoch, 275–294. Berlin: Springer.
Weber, Max. 1978. *Economy and Society,* 241–242. Berkeley: University of California Press [1922]; Shils, Edward. 1965. Charisma, Order, and Status. *American Sociological Review* 30(2): 199-213.
Wilf, Eitan. 2013.Toward an Anthropology of Computer-mediated, Algorithmic Forms of Sociality. *Current Anthropology* 54(6):716–739.

Chapter 6
Interstellar Travel Across Virtual Galaxies

A significant number and variety of massively multiplayer online games allow people to experience simulated interstellar travel in a social context, through first-person avatars, with varying degrees of realism. The most unrealistic feature they all share is that the time scale is vastly shorter than the real universe permits, often going from one solar system to another in minutes rather than centuries. Rather more realistic is the harvesting of extraterrestrial resources such as minerals or scientific data, moderately abstracted as simulations always are, but often conceptually rather complex. This chapter will concentrate on two very different but equally striking examples, *EVE Online* and *No Man's Sky*, the first well established and the second rather controversial. Brief surveys of science fiction in other media and of two other interstellar role-playing games, *Jumpgate* and *Elite: Dangerous*, will provide a conceptual background.

6.1 An Imaginary Launch

In 1865, Jules Verne published his influential novel, *From the Earth to the Moon*, imagining that humans could fly around our planet's satellite in a capsule fired from a gigantic gun.[1] That was the year in which the American Civil War ended, a conflict in which electronic communications in the form of telegraph had been extensively used by advancing armies and even by observers in balloons to tell their teammates on the ground what they could see of the enemy from far above the surface of the Earth. Verne's novel was somewhat realistic, except in minimizing the force of acceleration that would be required and the challenges of re-entry to the atmosphere and landing safely. Among the points of realism were that the capsule could not land on the Moon, and the duration of the journey was approximately correct.

In 1901, H. G. Wells used a more imaginary mode of transportation in *The First Men in the Moon*, that would allow astronauts to escape Earth at low acceleration

[1] Verne (1978).

W. S. Bainbridge, *Computer Simulations of Space Societies*, Space and Society,
https://doi.org/10.1007/978-3-319-90560-0_6

and land on the Moon (Wells 1967). His method was not entirely illogical, but did depend upon the existence of an otherwise benign substance, *cavorite*, which negates the force of gravity. Encasing the spacecraft with cavorite insulates it from Earth's attraction, while opening a window in the direction of the Moon allows its gravity to draw the astronauts to it. Of course, nothing like cavorite actually exists. In the abstract, this represents a standard tactic in science fiction of the subsequent years: Assume one counter-factual proposition that will allow interesting events to take place following logical principles in an otherwise normal universe. So long as the counter-factual is clearly stated, this can be an entirely valid method in computer simulations, admittedly giving results that are not entirely realistic, but functioning like an experimental manipulation in a laboratory study. If one unproven statement, X, were true, what would be the consequence, Y or Z?

The point at which science fiction shifted to rockets as the preferred mode of interplanetary transport was probably the 1929 movie *Frau im Mond* (*Woman in the Moon*), written and directed by Fritz Lang and presenting the multi-stage chemical liquid-fuel rocket ideas of Hermann Oberth who served as technical advisor (Oberth 1923). The one glaring inaccuracy was that the heroine referred to in the title did not need to wear a spacesuit while walking on the lunar surface, probably in deference to the men in the audience who needed to admire her beauty. This illustrates a general point that will be significant throughout the remaining chapters of this book: Human beings desire experiences that are not entirely compatible with the nature of the universe, requiring compromises in the design of simulations that may be instructive even as they are unrealistic.

One experience that was realistic and experientially desirable in all three of these examples was the duration of a flight to the Moon. Travel takes time. In reality, going to the Moon takes about the same number of hours as riding a railway train from New York to Los Angeles. There is nothing passengers need to do during these trips, except act out their personal dramas with each other. Aesthetically, a lunar voyage is rather like the middle movement of a typical concerto in classical music. The protagonist is like the featured pianist or violinist, holding center stage in interaction with the other performers. Since most concerti have three movements, the first represents the dramatic launch, the third is the dramatic landing, and the middle movement provides a period of rest and reflection that separates them.

A very few classic science fiction stories acknowledged that interstellar travel would take many years, notably "Universe" by Robert A. Heinlein, published in 1941 (Heinlein 1941). In his *Cities in Flight* series, James Blish suggested that voyage to another star would require two very radical innovations, one providing a new means for spaceship propulsion, and the other greatly extending human lifetimes (Blish 1970). A much more recent example, the 2017 movie *Passengers*, develops an interesting story around the long duration of interstellar travel, as described by Wikipedia:

> The starship *Avalon* is transporting over 5,000 colonists and crew in hibernation pods to the planet Homestead II, a journey that takes 120 years. 30 years into its journey, the ship encounters a meteor shower, in which it collides with a large asteroid. The ship's protective energy shield is heavily weakened as a result of the impact causing a malfunction that awakens

> one passenger, mechanical engineer Jim Preston (Chris Pratt), 90 years too early. After a year of isolation, with no company except Arthur (Michael Sheen), an android bartender, Jim contemplates suicide. One day he notices beautiful Aurora Lane (Jennifer Lawrence) in her pod. Her video profile reveals she is a writer with a humorous personality. After struggling with the morality of manually reviving Aurora for companionship, he awakens her, claiming her pod malfunctioned like his.[2]

This dramatic premise illustrates several relevant points. The somewhat realistic duration of the journey is clearly incompatible with ordinary human life, unless as in "Universe" and *Cities in Flight* the spaceship functions as a large, living community, at a high level of complexity, rather than being merely an interlude between adventures in different solar systems. Jim, the initial protagonist, develops relationships between two kinds of "people," neither of whom is identical to himself. Arthur is a robot, thus comparable to the non-player characters (NPCs) in role-playing computer games, simulated people who are largely pre-programmed but may possess a degree of artificial intelligence. Aurora is a real human being, of the opposite gender from Jim and thus biologically prepared to engage in a complementary human relationship, like many experienced by humans for the past million years, yet isolated from any wider society.

Much more commonly, science fiction follows the strategy of H. G. Wells, postulating a means of travel using poetic language that sounds like real science, but actually has no basis in fact. In his 1928 novel, *The Skylark of Space*, E. E. "Doc" Smith imagined a means of propulsion that would smoothly translate mass into velocity, without the inconvenience of pressure associated with acceleration or any barrier at the speed of light (Smith and Garby 1946). The 1934 novel *The Legion of Space* by Jack Williamson invented a device called *geodyne* that could warp space so that the spaceship could escape the ordinary geometry of the universe (Williamson 1947). By the 1940s when Isaac Asimov was writing his *Foundation* series set in an era after humanity had colonized the galaxy, science-fiction spaceships could jump into *hyper-space*, where either the laws of physics were different or the distances were much shorter (Asimov 1951). These literary conventions became popular in movies and television, notably in the *Star Trek*, *Star Wars*, and *Stargate* series, and thus were already well-established cultural elements when computer games began sending players to the stars.

6.2 Two Classic Galaxies

Online simulations of interplanetary and interstellar travel vary across many dimensions. To what extent is the environment organized around a story, comparable to a series of television programs but personally involving the player, versus being an open environment in which users decide for themselves what their goals will be? What is the balance between violent combat and more peaceful activities such as

[2]en.wikipedia.org/wiki/Passengers_(2016_film), accessed April 2017.

exploration and the economic exploitation of natural resources? What is the experiential perspective of the user, for example a first-person view from the cockpit of a spaceship and manually flying the craft in the manner of a fighter aircraft, versus third-person view looking at the spaceship from a distance and giving emotionally detached commands to an automatic guidance system? Two very clear examples, *Jumpgate* and *Elite: Dangerous*, will convey a sense of one of the classic approaches to simulating spaceflight.

Jumpgate was an online game that attempted to simulate propulsion of an interstellar spacecraft somewhat realistically, for example following Newtonian laws for acceleration. I explored it very briefly in May 2009, but it has since been shut down, and today the most effective way to study it may be through viewing videos players have posted on Internet, and by reading reviews. The primary mode of experience and control is first-person, flying the spaceship like a fighter jet, but there is also a map-focused view in which the pilot can set a chain of steps toward a destination.

Launched in 2001, *Jumpgate* depicts a future time when galactic disaster has caused social disorganization, and factions are competing to reconstruct interstellar civilization. As an IGN review explains, there are several ways in which this is a social simulation: "As you play the game you not only fly and fight, but you team up with others to actually re-create a universe with all that such an enterprise entails. You will interact with squadron members and wayward travelers, struggle to stay on course in the face of solar storms, and battle mysterious human and alien adversaries."[3]

Finding YouTube videos that depict a computer game from the standpoint of the player is rather easy, and the challenge often is that too many of low quality are available, swamping those in a keyword search that would be especially valuable for the researcher. One strategy is to sort the videos by the number of times each has been viewed, but this is less than ideal when one's research interests are at all specialized, and when the videos were uploaded on a wide range of dates. An additional challenge for *Jumpgate* is that variants of it have repeatedly been made available from different legal and illegal providers, typically lacking full documentation of the vintage of the version or when the video was taken rather than uploaded. A good starting point in this case is the remarkable set of videos of many games uploaded by someone calling himself Space Game Junkie, especially valuable because he provides extensive displays of a large number of space-related computer games.[4]

For this chapter, a video "Trying to Find Pure Asteroids in Jumpgate," specifically dated to November 27, 2015, and viewed only 162 times by other people, is especially useful.[5] It was live streamed for 2 h and 37 min, thus realistically depicting actual play, with a realtime narrative by the player. It is social in two ways. First, mining asteroids is one role in a complex economic system, involving other players in other roles. Second, the very posting and viewing of YouTube videos is a form of communication

[3]www.ign.com/games/jumpgate-the-reconstruction-initiative/pc-16696, accessed March 2017.

[4]www.youtube.com/user/SpaceGameJunkie/playlists?view=1&sort=dd&shelf_id=3, accessed April 2017.

[5]www.youtube.com/watch?v=RsWfTZUSBgA, accessed March 2017.

between people, which in cases like this involves sharing their experiences interacting with a computer simulation of outer space.

For readers who have not viewed videos of this type, it should be noted that they are often highly informal, including emotional outbursts and ambient noises, but in this case giving an exceedingly informative introduction to the human as well as technological meaning of *Jumpgate*. The narrator mentions in passing that he has downloaded the game from a Russian site, presumably created by someone who had obtained or reverse engineered the software operating the database and Internet server, presumably also in violation of copyright, but unlikely to suffer legal reprisals because the company that owned *Jumpgate* no longer existed.

The narrator was an experienced *Jumpgate* player, claiming to have played more than 2,000 h in earlier years, but returning for a festival scheduled for the Thanksgiving weekend, and thus having to remind himself about some aspects of the game. Something unexpected happened during his session, so he later posted a blog which I found by googling some of the terms he used in the video, rather than directly through a link from YouTube. One event in the festival involved competition to mine more "pure asteroids" than anybody else, and his blog said:

> To explain real quick, 99% of the asteroids in Jumpgate are static, both in position and type (and there are multiple types), and always vary in their yields of specific commodities like ice, precious metals and the like. Pure asteroids, however, pop up randomly within a certain radius of the gate at the center of the system (the o.o.o coordinates) and are finite, meaning they eventually pop, and another random asteroid takes their place on the map somewhere. For me, it's a terribly fun little hunt that takes patience and skill, and it's one of the most enjoyable things in the game for me (Rubin 2017).

To mine an asteroid, one must find it within a particular solar system, approach it with a spaceship that carries a mining laser, then hold the ship steady while firing the laser. This is a common device in several space games, suggesting a plausible technology in which a laser vaporizes and ionizes the material of the asteroid, then attracts some portion of the residue into the hold of the spaceship. Ordinary asteroids in *Jumpgate* continue to exist after a mineral has been completely extracted from them, while a pure asteroid is totally destroyed, what he means by the word "pop." The mining task takes the last hour of his video, while the first hour and a half are devoted to delivery missions and an unexpected encounter with an alien intelligence.

His spaceship begins a great distance from any worthy destinations, and we learn how to navigate by watching the first few minutes. A pilot opens a map of the galaxy, in the form of a network of jumpgates, large structures floating in space that can teleport a spaceship to a comparable jumpgate in another solar system. The pilot marks the chain of destinations on this map, then closes it. All the subsequent spaceflight action is seen from the pilot's seat in the cockpit, and the ship must be flown manually to the first jumpgate, then after jumping through it to the next jumpgate, and so on. In this episode, he stops at his home base to buy "gravitational components," which he plans to sell for a profit at his next destination, which he reaches one full hour after the beginning of the video.

To undertake a space mission that is especially appropriate for the festival, he buys many bottles of "Solrain Stout," a cosmic brand of beer, and heads back by a different route toward his home solar system. Suddenly, in a solar system called The Dark, he encounters a sentient entity called a Conflux, which begins to chase him. He has filled his ship with cargo, more than just the beer, so he cannot outrun this hostile alien, and expects to die any moment. Then, to his extreme surprise, it communicates with him, and he realizes that it is actually operated by a GM—a Game Manager—some Russian computer expert having control over the game. Safely reaching his home system he sells the beer and other cargo, then goes in search of a solar system that contains a pure asteroid.

The first system he checks, just one jump from home, contains only an ordinary asteroid, but the second solar system does have a pure one. He fires his mining laser at it, and outside the front window of the cockpit we see a beam of fire from the ship striking the asteroid. It is composed of pure germanium, and as soon as it has been exhausted, a second pure asteroid appears nearby, in sequence until the fifth fills the cargo hold, now containing also phosphorus, erbium, antimony and chromium. The mission is completed when he sells these elements at the market in his home system.

The social aspects of this two and a half hour adventure are somewhat limited, most intense while interacting with the GM masquerading as a Conflux. There seem to be only six other players online, and one person viewing the video during its live stream, but he communicates actively with the former by text chat, and the latter by voice. At two separated space stations he buys and sells resources, thus connecting economically but indirectly with other players. It is often difficult for researchers to know what fraction of such a virtual economy is pure trade between players, versus also including automatic buying and selling by the database, but certainly at least part of the market is created through human action. Despite its obvious limitations, this videoed adventure and commercial activity do provide a good picture of fundamental principles found in much more populous simulations of galactic societies.

In a 2001 review on GameSpot, Ron Dulin compared *Jumpgate* with several existing games: "Jumpgate is more than just an online space combat game; it successfully incorporates the trading elements of Elite and Privateer, the combat elements of Freespace and Wing Commander, a lighter version of Independence War's realistic physics, and all the level-gaining and experience bonuses that make online role-playing games so addictive (Dulin 2017)."[6] The reference to *Elite* concerns the series of galactic simulations mentioned in the first chapter of this book, currently represented by *Elite: Dangerous*. As I did in 2009 with *Jumpgate*, I very briefly entered *Elite: Dangerous* in 2017, finding many similarities with *Jumpgate*.

Figure 6.1 is a scene taken from one of the beginner tutorials in *Elite: Dangerous*, in which a novice player learns how to fly across a solar system to an orbiting space city and dock with it. The view is through the pilot's cockpit window, with the main interface display across the bottom, and four windows separated by dark frames. The complex, obscure shapes at the top are the orbiting space city, and the gray shape on the dashboard at the bottom and left of center is how the control interface depicts it.

[6]Dulin, Ron. 2017. Jumpgate review. *GameSpot,* October 19, 2001, http://www.gamespot.com/reviews/jumpgate-review/1900-2819087, accessed March 2017

Fig. 6.1 The cockpit view of an orbiting city and its planet

The ship is pointed between the city and the planet it orbits, with the main band of the galaxy's disk behind the planet and up to the right.

As explained in Chap. 1, a fundamental principle of the *Elite* series, progressively realized at increasing levels of complexity, is procedural generation, a computational method to simulate a very large number of planets and solar systems. Wikipedia reports: "Piloting a spaceship, the player explores a realistic 1:1 scale open world galaxy based on the real Milky Way, with the gameplay being open-ended. The game is the first in the series to attempt to feature massively multiplayer gameplay, with players' actions affecting the narrative story of the game's persistent universe, while also retaining single player options."[7] An independent wiki created by the community of players says *Elite: Dangerous* was programmed "to simulate a realistic Milky Way galaxy with realistic star systems and seamless freeform interplanetary and atmospheric flight and landing on moving planets, with realistic orbits and day/night cycles. The developers have gone to great lengths to map and electronically recreate the Milky Way to scale to make the game as realistic as possible."[8]

A key factor in understanding the four games described in this chapter is that they were created by small groups of creative computer enthusiasts, *Jumpgate* in Colorado and the other three outside the nations with the biggest videogame industries, which are the United States and three Asian countries: Japan, China and South Korea. Like the *Elite* series, *No Man's Sky* was created by a small group in Britain. As Alison Gazzard has reported, the original *Elite* game was specifically designed for the British BBC Micro educational personal computer, and today's *Elite: Dangerous*

[7] http://en.wikipedia.org/wiki/Elite:_Dangerous, accessed December 2017.

[8] http://elite-dangerous.wikia.com/wiki/Elite_Dangerous_FAQ, accessed August 2017.

is to a significant extent the result of cultural nostalgia (Gazzard 2013).[9] It would be too much to claim that *Elite* was an attempt to resurrect the British Empire in a virtual context, and *EVE Online* originated in Iceland, a different island nation. Yet there is some truth to the metaphor that simulations of galactic colonization allow people who are currently unable to expand their scope across our own globe, to experience a virtual alternative.

6.3 A Galaxy of Algorithms

By far the most impressive and influential interstellar role-playing simulation is *EVE Online*, which launched in 2003. The author has already published two chapters about EVE, the more recent of which viewed the galaxy from the perspective of an avatar based on rocket pioneer Hermann Oberth, and was published in the 2016 collection by many authors, *Internet Spaceships are Serious Business: An EVE Online Reader*.[10] The concluding chapter of that anthology, written by five archivists, reports that EVE has been honored as a cultural and technological landmark: "In fall 2012, the Museum of Modern Art's (MoMA's) Department of Architecture and Design (A&D) announced its intention to officially acquire video games for the permanent collection. One of the first fourteen titles to go on exhibit was *EVE Online*, making it the first massively multiplayer online role-playing game (MMORPG) to be acquired by a museum (MacDonough et al. 2016)."

Created and managed by a small team in Reykjavík, Iceland, EVE is a prime example of the social origin of many computer games in small groups of independent enthusiasts, who faced extreme challenges bringing EVE to market and building its player base. Their Wikipedia page describes the organization: "CCP hf or CCP Games (Crowd Control Productions) is an Icelandic video game developer and publisher, majority owned by the company's staff and founders, Novator Partners and the American investment fund General Catalyst Partners."[11] In a 2005 interview, senior EVE producer Nathan Richardsson provided some of the cultural and psychological background: "The founders had two passions which they wanted to join, the sci-fi feel and vastness of space from Elite and the social interaction of massively multiplayer and player vs. player gaming from Ultima Online. I should also add that they were quite active PvPers in UO and this is the main reason for our emphasis on PvP. We feel that the emotions involved with losing something of value is just as important as gaining something of value, it makes a very immersive experience. There have

[9]Gazzard, Alison. 2013. The Platform and the Player: Exploring the (Hi)stories of Elite. *Game Studies*, 13(2), http://gamestudies.org/1302/articles/agazzard

[10]Bainbridge (2011), Carter et al. (2016).

[11]http://en.wikipedia.org/wiki/CCP_Games, accessed March 2017.

to be lows to make the highs more enjoyable. PvP allows us to achieve that."[12] The following chapter will examine more closely the historic role of *Ultima Online*.

Set 21,000 years in the future, *EVE Online* postulates that spacefaring humanity had discovered a wormhole they named EVE, through which they could begin to colonize a distant galaxy they named New Eden. But the wormhole unexpectedly closed, and the colonists were not yet fully self-sufficient, causing a near collapse of the colonies. Gradually, four different civilizations arose from their wreckage, each with a somewhat different culture: the theocratic Amarr, the inventive Minmatar, the capitalistic Caldari, and the sensuous Gallente.[13] When the player enters this universe, the four interstellar civilizations compete but have established a slightly shaky peace across some solar systems, leaving most of the galaxy in anarchy. The player's avatar initially belongs to one of the four civilizations, and learns how to function in regions, called *high-sec* for *high-security*, protected by non-player-character police called CONCORD. The margins of civilization are *low-sec*, and in much of the galaxy, *null-sec* regions, there is no law or order unless administered by the players.

The result is that EVE simulates not merely spaceflight but the realistic development of an entire complex of societies, born in the chaos of war but increasingly organized. This often leads to very large-scale conflicts, notably a March 2016 war involving 60,000 players, with about 4,300 participating in the biggest single battle in which each player operated a spaceship. Leading EVE researcher Marcus Carter reports:

> A war in EVE is humbling. The largest Alliances have tens of thousands of players, and a war can involve two to three battles a day for weeks. Complex military command structures are established, dictating broader strategies and war theaters. Diplomatic efforts run parallel, cajoling, bribing, or threatening other Alliances into supporting (or keeping distance from) a war. Attacks are planned and led by 'fleet commanders' who lead hundreds, and occasionally thousands, of players into battle, flying ships that accord with carefully theorycrafted fleet doctrines. A single battle can see ships worth the equivalent of US$300,000 permanently destroyed. Teams of players ensure local in-game markets are stocked with these ships, and control over the minerals to build them enters the strategy of wars (Carter 2014).

The game's website advertises the options concerning social organization: "Player-created empires, player-driven markets, and endless ways to embark on your personal sci-fi adventure. Conspire with thousands of others to bring the galaxy to its knees, or go it alone and carve your own niche in the massive EVE universe. Harvest, mine, manufacture or play the market. Travel whatever path you choose in the ultimate universe of boundless opportunity. The choice is yours in EVE Online."[14] A prime example of a benevolent organization is EVE University, which describes itself thus on its wiki: "EVE University is a corporation in EVE Online and a member of the Ivy League Alliance. We are a neutral, non-profit training corporation in New Eden.

---

[12]Quoted by Jim Rossignol, "Interview: Evolution and Risk: CCP on the Freedoms of EVE Online," *Gamasutra*, September 23, 2005, www.gamasutra.com/view/feature/2411/interview_evolution_and_risk_ccp_.php.

[13]http://eve.wikia.com/wiki/Races_and_Bloodlines, accessed March 2017.

[14]www.eveonline.com, accessed March 2017.

Founded in March of 2004 by Morning Maniac, EVE University continues to take new pilots and train them in all aspects of EVE Online. As well as the Uniwiki, EVE University runs classes on various subjects, the majority of which are open to the public both live and in our Class Library."[15]

This simulated universe is not only technically complex, but simulates serious labor, as a team of experienced researchers reports: "EVE is not simply *like* work for some players and in some formats, but is *already work*—that is, activity that generates wealth within and for a broader economic order that is itself increasingly virtual (Taylor et al. 2015)." EVE is said to possess a "punishing learning curve" and to be "extremely difficult to learn to play (Paul 2011; Bergstrom 2012)." But much of the action in EVE is technological, for example harvesting metals from asteroids in order to build complex new spaceships—miners and freighters as well as warships.

In my first phase of research, 2009–2010, I ran four characters a short distance, one through each of the four civilizations, to get a sense of the background lore. In 2013, when each account could have a maximum of three avatars, I retained two of the original avatars and added a new one with the specific research goal of understanding the space-related technologies. Returning again in 2017 to write this chapter section, I have these three: (1) Theo Logian in the Amarr, who is a devout practitioner of his culture's religion, (2) Cogni Tion, a scientist among the Minmatar, and (3) Herman Noberth, an engineer and spaceflight visionary, also among the Minmatar. It is noteworthy, that I could not use the exact name of real spaceflight pioneer, Hermann Oberth, because somebody else had already used it, and avatars must have unique names. In a chapter of the EVE anthology, I used Noberth to explore spaceship technology, especially investing time constructing a larger number of different spaceships, some designed for battle and others for commerce (Bainbridge 2011; Carter 2016). Here, Noberth will explore the depths of mining, including the construction of the necessary specialized equipment.

Figure 6.2 shows Noberth's mining barge, of the design named Procurer, working in an asteroid belt in the Stirht solar system. The bright beam shooting from the turret atop his ship to the target asteroid is the laser from a Strip Miner 1 device. It progressively vaporizes the surface of the asteroid, and the material is magnetically drawn into the ship's ore hold. Five practically invisible light beams around the asteroid are from drones that Noberth deployed. They use the same process to extract the mineral, and periodically return to the ship to unload. When the ship's ore hold is full, or the asteroid has been destroyed, Noberth will need to instruct the drones to return to their bay of the ship. Then he either takes the mined mineral to a space station for sale or processing, or if the ship still has room, he starts the process over again on another nearby asteroid.

To mine an asteroid in a particular solar system, Noberth would begin with his ship floating in space inside that system, and open a data list to see the mining opportunities. The geometry of asteroid belts is rather different in the EVE galaxy than our own. Each asteroid belt is an arc or circle of asteroids less than 50 km across,

[15]wiki.eveuniversity.org/Main_Page, accessed March 2017.

Fig. 6.2 A spaceship using a laser to mine minerals from an asteroid

associated with a particular planet, and some planets have several. The following steps start the mining process:

1. Select asteroid belt
2. Make ship warp to within 0 m of the belt's center
3. Select an asteroid by locking it as the target
4. Start an approach to the asteroid
5. Within 15 km start Strip Miner 1
6. Within 10 km launch 5 mining drones
7. Set drones to mine repeatedly
8. Stop ship and hover, or orbit the asteroid, waiting for the mining to complete.

The center of an asteroid belt is always empty, and the warp sends the ship quickly to that location, usually around 20 km from many of the asteroids. The database lists them in terms of ore that can be obtained, which for asteroids in the high-security regions is usually plagioclase, scordite, and veldspar. The asteroids differ in quality, and Noberth preferred those composed of rich plagioclase ore. A German website offering analysis of changing prices for minerals incidentally lists which minerals can be extracted in which proportions from specific kinds of ore.[16] When veldspar is processed, it produces a mineral called tritanium—not to be confused with the terrestrial chemical element, titanium. Scordite produces tritanium and pyerite. Plagioclase produces tritanium, pyerite, and mexallon. The standard way to do the processing is to visit a space station having the correct facility, and pay a small fee.

Once one has obtained some minerals, they can be used to manufacture useful tools, such as the strip miner and mining drones with which Norbert did much of

[16]http://ore.cerlestes.de/#site:ore, accessed March 2017.

Table 6.1 Statistics about the requirements for manufacturing mining equipment

	Miner 1	Strip miner	Mining drone 1	Ice mining laser	Ice harvester	Ice harvesting drone
Blueprint ISK	92,720	16,130,000	19,986,000	3,500,000	15,003,680	30,000,000
Mineral units:						
Tritanium (4 ISK)	1,324	24,663	77	8,095	30,710	385
Pyerite (5 ISK)	481	19,132	0	4,120	17,250	0
Isogen (50 ISK)	0	0	4	1,755	1,527	20
Mexallon (65 ISK)	119	7,806	0	2,025	4,921	0
Nocxium (384 ISK)	0	1,351	13	125	597	65
Zydrine (1,120 ISK)	0	0	0	0	300	0
Megacyte (1,490 ISK)	0	528	1	0	512	2
Morphite (11,000 ISK)	0	0	0	0	0	0
Total Mineral Cost ISK	15,436	2,007,206	6,990	320,355	1,933,433	30,480

his ordinary mining. Also required is the correct blueprint, which must usually be purchased, and training to the correct level in the necessary skills, an activity that will be described later. Table 6.1 provides information about the blueprint costs and mineral supplies needed to make each of six mining tools, three for ore and three for ice, all of which Noberth did in fact produce. A single load of 12,000 cubic meters of rich plagioclase ore can be sold in the Stirht solar system for around 2,000,000 ISK, only a small fraction of the cost of the six blueprints. This currency unit has the same abbreviation as the Icelandic króna, the currency of EVE's home country, but in EVE actually stands for "Inter Stellar Kredits." One blueprint may be used to make an endless number of items, and Miner 1 requires only the minerals that can be obtained from plagioclase. However, this tool mines very much more slowly than a strip miner, let alone a strip miner supplemented by five mining drones, so Miner 1 is more a training tool for beginner than a valuable tool for advanced miners.

While I was mining in the asteroid belt of the innermost planet of the Stirht system on day, I noticed there were two wrecked spaceships in the vicinity, a Gallente destroyer and a Minmatar frigate. My first guess was that this was a mere simulation of the residue of a space battle, given that it was in a high-security system. However, as Richard Page had noted in an essay of how deviousness was valued in EVE, teams of players do sometimes enter high-sec space with the goal of *ganking* weaker players, even if this means losing some of their own ships to the Concord police (Page 2016). Wikipedia defines this as one of the modus operandi of PvP: "Ganking (short for gang killing) is a type of Pking [Player Killing] in which the killer has a significant advantage over his victim, such as being part of a group, being a higher level, or attacking the victim while they are at low health."[17]

Had the two wrecked ships simply battled each other, the Gallente destroyer would have almost certainly defeated the Minmatar frigate, implying that more than these two ships would have been involved. The simplest scenario was that the destroyer vanquished the frigate but was then smashed by police before it could escape. So, this was apparently a case of suicide ganking, as explained by the wiki of EVE University: "As CONCORD will always react to destroy any ship which acts in such an aggressive manner, the aggressor is guaranteed to lose their ship—hence the term 'suicide'. However, CONCORD does not react instantly, giving the attacker time to try and destroy their target. The aim of suicide ganking therefore is to destroy a higher value target than the value of the ship being used to gank."[18]

I checked an online database of EVE battles a couple of days later, after there had been time for the episode to be reported, and indeed this had been a real attack, earning the winner 11,130,000 ISK.[19] The victor in that local battle had been Mechjeb Kerman, a member of the Horde Vanguard corporation, who had piloted the Gallente ship. Over his illustrious career, he had destroyed fully 3,250 ships of other players, earning 684,81,000,000 ISK at a cost of losing 799 of his own ships. Horde Vanguard had 154 members, who collectively had destroyed 25,364 ships, earning 4,530,000,000,000 ISK with a loss of 4,425 of their own ships. The Minmatar ship belonged to a simulated group of non-player characters called the Thukker Tribe.[20] So this was not a case of player-versus-player murder.

To get a sense of the wider context, we can compare mining with salvaging, which can extract resources from shipwrecks. I checked the local markets and discovered that salvage drones could be purchased for 27,000 ISK each, so I bought 5 and then began the training necessary to use them, aware that the wreckage would be gone before I could salvage it, but wanting to prepare for future opportunities. First, one must get a general skill named Survey up to level 3 on its expertise ladder. That requires a skill book, which cost 36,000 ISK in this case, and putting it into the training queue of the avatar. This is the way skill training works in general. One buys a training book, which contains lessons for the 5 levels in that skill. The skill queue is a list of the order in which the avatar will learn the next few lessons. Level 1 for Survey took just 8 min, but level 2 took 39 min, and level 3 took 3 h and 40 min. Then one is able to learn Salvaging, the book for which cost 1,000,000 ISK. Level 1 of Salvaging went into the skill queue even before level 1 of Survey had been learned,

[17]http://en.wikipedia.org/wiki/Player_versus_player#Player_killing, accessed March 2017.

[18]wiki.eveuniversity.org/Suicide_ganking, accessed April 2017.

[19]zkillboard.com/system/30003376/, accessed March 2017.

[20]eve.wikia.com/wiki/Thukker_Tribe, accessed April 2017.

and it would require 25 min after level 3 of Survey had been gained. The next item in the queue was Salvaging 2 which took 1 h and 57 min. The book for Salvage Drone Operation was rather cheaper, only 220,000 ISK, and its first 3 levels would take 34 min, 2 h and 35 min, and 14 h and 38 min.

Skill training continues in the background, even after one logs off and shuts down the computer. Obviously, the routine in the program that does the training merely checks the computer's clock, and does whatever updates to the skill database are indicated, rather than employing some real process of intellectual enhancement gradually taking place. Thus it is simulated learning. The higher a level, the longer it takes to train, and level 5 of Salvage Drone Operation would take 19 days, 12 h, and 20 min. At this point, I had already completed level 5 in both Drones, Mining, and Mining Drone Operation. To get started on salvage, I had halted training the fifth level of Mining Barge, which improved the efficiency of mining using the Procurer, and had 3 days and 8 h left before completion.

Once all the devices listed in Table 6.1 had been produced, it was time for Noberth to leave his home region in Minmatar space, explore high-sec regions of the other factions, and mine a wider range of materials. He began in the Aminaka solar system where he had done most of his manufacture and set a course for the home system of the Amarr, which the EVE University wiki reported was one of 5 main commercial hubs in the galaxy, two of them being Minmatar, in the Rens and Hek solar systems. So he flew to the Brutor Tribe Treasury in orbit around Moon 8 of the Rens VI planet, from there to the Boundless Creation Factory in orbit around Moon 12 of the Hek VIII planet, then through Gallente and Caldari space without visiting their trade hubs, to Emperor Family Academy in orbit of Amarr VIII. The EVE University wiki explains that the Amarr hub is the second richest, a total market offers value of 693.3 billion ISK, compared with only 137.9 billion ISK at Rens, but that both pale in comparison with Jita in space controlled by the capitalist Caldari, with a total value of an astounding 119.4 trillion ISK:

> Jita is the center of trade in New Eden and is by far the largest hub in game, often having over 1000 players in system with consistent traffic in and out. Trading in Jita is extremely competitive due to the high level of activity. This often results in fairly low profit margins, and is harder for players with low ISK to succeed in due to the richer able to shrug off the low margins by trading in volume. However, players new to trading will be able to turn over smaller orders more quickly, and therefore learn what to do/what not to do sooner. Be mindful of low volumes as well, since if you do find products with decent profit margins (say above 20%), these items flow a lot slower than those with narrower margins.[21]

Along the way, Noberth was able to mine and refine both isogen and nocxium, and salvage wreckage from what appeared to be another suicide gank. In 31 solar systems he was able to locate fully 323 asteroid belts, but saw no hint of ice. To find ice, he needed to use special equipment to scan the local solar system. An online maps system told him that two systems very near Amarr, Esteban and Warouh, often contained unstable ice fields. He set up his Procurer with 5 salvage drones and 5 mining drones in the drone bay of the ship that could carry exactly ten of these types,

[21] wiki.eveuniversity.org/Trade_Hubs, accessed April 2017.

Fig. 6.3 Three mining ships working in the same asteroid field

an ice harvester in one of the ship's two high energy fixtures and a scanning probe launcher in the other. He temporarily placed his beloved strip miner in the ship's regular hold, which he had expanded with two additional cargo holds, along with backup probes and the materials he salvaged from wrecked ships. Figure 6.3 shows his Procurer in the center of the image, firing the ice harvester out of view to the lower left, as two other ships work in the background.

The left side of the image shows a player-operated exhumer ship of the type named Mackinaw which can hold 35,000 cubic meters of ore, compared with the mere 12,000 cubic meter's of a Procurer. Noberth's ship is much nearer and is somewhat smaller than the Mackinaw. The ship to the right is an Orca freighter, vastly larger than either of the mining ships. Noberth had seen the pair earlier in an ordinary asteroid belt, and clearly the freighter was being used to store vastly more ice, and deliver it for sales or use in another solar system, perhaps some distance away. The database identified the two avatars that operated the ships, but we do not know if actually they represented cooperation between two players, because some players use two accounts simultaneously.

A section of my earlier book, *The Meaning and Value of Spaceflight*, explored what simulated space activities might mean to EVE players, by tabulating from February through July 2013 the goals listed in the in-game advertisements of 553 of the formal groups called corporations (Bainbridge 2015). An avatar can belong to only one of these groups, but some of them cluster into larger alliances, and given EVE's focus on large-scale conflict, relations between them can be exceedingly complex. For example, a player may use a secondary avatar to infiltrate a group that is the enemy of the group to which that player's main avatar belongs. Here, we can use mining as the entry point to a brief analysis of more peaceful social relations, analyzing the data in a new way, through correlation coefficients.

Of the 553 player groups, 165 or 29.8%, listed mining among their goals. A crosstabulation of groups versus goals was created, in which a "1" indicated that the group practiced mining, while "0" indicated that it did not. The correlation between mining and the membership size of the group is -0.17, indicating a small tendency for large groups to ignore mining. In terms of average membership size, the difference is actually rather large. The average size of groups that list mining among their goals is 18.8 members, compared with 35.6 members for groups that do not mention mining. This naturally raises the question of how goals cluster across the groups, most specifically which other goals correlate with mining, and how those goals relate to group size. The results of this correlational analysis in Table 6.2 will not be especially surprising, but they do illustrate how patterns of empirical data can be extracted from virtual world simulations involving human beings.

As we might expect, mining groups tend also to manufacture goods and sell them. In EVE, the technical term *research* refers to technical action related to manufacturing that improves the quality of a blueprint, so that its particular product can be produced more cheaply and quickly. Notice that the numbers of groups advertising these four goals differ significantly: mining (165 groups), manufacturing (103), trade (97), and research (81). Given the cost of blueprints, manufacturers tend to specialize, and large combat-oriented corporations that do not list manufacturing as a goal may include a few players with avatars devoted to manufacturing spacecraft and their accessories. Corporations often cluster in larger alliances, and may trade informally within the alliance rather than using the public marketplace.

Clearly, the social and economic structure of *EVE Online* is exceedingly complex, and "research" is given a rather narrow meaning. We can now consider a very different universe in which social interaction is problematic, and "research" takes on much more central significance.

6.4 Asocial Social Exploration

In the summer of 2016, great excitement for a galaxy being created by an independent British game developer, Hello Games, swept our local World Wide Web. The home page for *No Man's Sky* says it "is a game about exploration and survival in an infinite procedurally generated galaxy" but does not fully explain the extent to which it is a social rather than merely astronomical simulation: "Choose whether to share your discoveries with other players. They're exploring the same vast universe in parallel; perhaps you'll make your mark on their worlds as well as your own."[22] Prior to its launch August 9, 2016, many gamers believed it was like *EVE Online*, a massively multiplayer role-playing game with extensive social interaction. Two of the most prominent social game blogsites, Massively and MMORPG.com, distributed pre-

[22] www.nomanssky.com/about, accessed April 2017.

Table 6.2 Correlations connecting goals advertised by 553 corporations in *EVE Online*

Goal	Description	Correlation with mining	Correlation with size	Groups
Manufacturing	Assembling components and raw materials to build a wide variety of products, requiring expensive blueprints and training that takes many days but can be performed by an avatar while the player is offline	0.66	−0.09	103
Trade	Buying and selling loot and manufactured goods, through a market distributed across the civilized solar systems	0.62	−0.17	97
Research	Improving "blueprints" used to manufacture machinery, not to improve the product's performance but to reduce the cost in raw materials required to produce it	0.58	−0.09	81
New-pilot friendly	The corporations that advertise especially aim their messages at new players, who generally finish many pre-determined tutorials before joining a corporation	0.31	−0.05	297
Mission running	Performing pre-scripted story-based activities, assigned by non-player characters belonging to the four societies or other fictional organizations	0.13	−0.10	254
Roleplay	MMOs are generally role-playing games, but this aspect is less significant in EVE after the early pre-scripted missions	0.12	−0.08	74
Exploration	Finding concealed sites or objects in space	−0.02	−0.01	221
Bounty hunting	Attacking ships belonging to wanted criminals	−0.06	−0.06	64
Factional warfare	Open conflict among four interstellar societies (Amarr, Minmatar, Caldari, Gallente), with the player's corporation belonging to one	−0.10	−0.01	93
Incursion	Skirmishes against invasions by non-player spaceships	−0.12	−0.07	149
Alliance warfare	Conflict between groups of corporations allied with each other	−0.16	0.17	139
Piracy	Attacking other players for the loot gained if their ship is destroyed, or to force them to pay ransoms	−0.23	0.02	143
Small-scale gangs	Combat between small groups of players, often lacking much experience, that can prepare them for large-scale warfare between large corporations later on	−0.28	0.13	269

launch publicity, reinforcing this false assumption. On August 10, chief Massively editor Bree Royce called it "the non-multiplayer multiplayer game" and reported:

> If you were confused about just how multiplayer No Man's Sky was going to be, welcome to the club. Inconsistent (or consistently contradictory) statements from Hello Games on the topic in the long lead up to launch didn't help. But since the game's launch yesterday, players have put it to the test, and we are… not really much closer to the truth. Kotaku reports on one pair of players who found themselves on the same planet and streamed their attempts to meet up, though it ended in failure. That prompted Hello Games' Sean Murray to tweet that multiplayer (in the game he said was not multiplayer two days ago) is possible, has been anticipated, and will be encouraged.[23]

A week later, William Murphy wrote on MMORPG.com: "No Man's Sky is a hard game to quantify. I mean, we initially included it on MMORPG's list because we (along with many others) expected it to be a multiplayer experience akin to say Elite: Dangerous. But as launch drew closer, we began to realize that No Man's Sky is indeed a single-player experience."[24] I had begun to explore NMS on August 12 and quickly discovered that its universe was so vast that its was indeed unlikely that I would encounter another player by chance. The only interaction with other players apparently built into the simulation was that players could upload names for the planets and animals they discovered. There was, however, a way that players could interact much more vigorously, but in our own galaxy, namely by joining Facebook groups. Table 6.3 lists the groups I studied, having to join the closed groups to see their postings, with their numbers of members and administrators as of March 18, 2017.

Another closed group, Sky Quest—No Man's Sky, had proclaimed: "Our goal is to find another member of SQ in this nearly infinite game, and to defeat the odds of never finding other intelligent human lifeforms." But when I contacted the founder in March 2017 I learned that the group had become inactive, apparently having failed to achieve its goal. My own goal in entering NMS was simply to gain enough first-hand experience, extensively exploring only six planets in three solar systems, to make sense of the communications that a great diversity of more active explorers were sharing in the Facebook groups. To provide enough background for the reader, we shall begin with documentation of two species of animals on different worlds.

Figure 6.4 shows a gigantic but docile vegetarian animal called *ocolrumgr brunicu*, weighing fully 261 kg, on the planet Mookamoganeil, in a solar system named Pujarafnsoka, where I also explored the planet Wayddiomitol. These are the randomly-generated names the animal and planets already had when I encountered them, and I chose not to invent new names for them. The image also includes large plants, some of which have very large spherical bases. The wildlife on Mookamoganeil is diverse, including small creatures that run quickly away, and hostile predators of dangerous size.

---

[23] massivelyop.com/2016/08/10/no-mans-sky-the-non-multiplayer-multiplayer-game, accessed April 9, 2017.

[24] www.mmorpg.com/no-mans-sky/reviews/for-the-un-hyped-an-exploratory-experience-pc-review-1000000451, April 9, 2017.

Table 6.3 Facebook groups oriented toward *No Man's Sky*

Name	Type	Founded	Members	Admins
No man's sky	Closed	December 23, 2014	18,978	12
No man's sky community	Closed	May 18, 2015	14,674	13
Explore no man's sky	Closed	October 17, 2015	3,260	10
No man's sky galactic survivors	Closed	April 11, 2016	1,382	5
no man's sky: atlas	Public	March 30, 2016	1,046	6
NMS love	Closed	August 23, 2016	841	5
No mans sky PS4	Closed	July 11, 2016	686	3
No man's sky fans only	Closed	August 18, 2016	550	4
No mans sky youtubers	Public	June 2, 2016	343	3
No man's sky community (let's explore)	Public	April 29, 2016	211	1
No man's sky	Public	January 25, 2015	174	4
No man's sky photography	Public	March 11, 2017	98	4
T.G.S no mans sky vblog page	Public	October 22, 2015	61	3
No man's sky fanatics	Closed	March 20, 2016	58	2
No man's sky bounty hunters	Public	August 8, 2016	44	1
No man's sky galactic family	Public	September 10, 2016	38	1
No man's sky Ps4	Public	August 8, 2016	38	1
No mans sky community	Public	August 10, 2016	36	1
No man's sky discoverys	Public	July 31, 2016	33	1
No man's sky	Public	May 15, 2016	30	1
No man's sky fans	Pubic	August 30, 2016	25	2

Among the pre-set goals on each planet, which earn rewards if accomplished, are discovering six locations where bases or beacons remain from earlier habitation, including some that may contain an intelligent extraterrestrial, and scanning a variable number of animal species for information about them. In an early blog about how to discover new species, Richard Scott–Jones wrote that simulated equipment called an analysis visor could be very useful:

> Grey pulsating dots indicate species that your visor can detect, but which are too far away to be identified. When you get close enough, light green dots indicate creatures you've already scanned, while red dots are unknown. Creatures can look similar, but be different. Sean Murray has said that code exists in No Man's Sky which simulates evolution, meaning creatures in the game may develop similar body parts as they adapt to a particular planet. We love this aspect of the game for its realism, but it can be a little misleading when trying to catalogue the species on each planet. The lesson: just because this squirrel-like thing looks similar to the squirrel-like thing you already scanned, don't assume it is. It could have a

Fig. 6.4 An ocolrumgr brunicu on Mookamoganeil in the Pujarafnsoka solar system

> different tail, or something, and though these differences may be minor, they all still count as individual analyses. Take the extra second and scan everything (Scott-Jones 2017).[25]

The Reddit discussion website has a subreddit named NoMansSkyTheGame with 129,025 subscribers, where an early posting by Kanuhduh triggered a frustration-filled chat about how to earn the bonus for discovering the last of the required animal species on a planet. The NMS user interface has a discoveries component with a page for each planet, listing among other things the discovered species plus check-box-marked lines for the number not yet found. Kanuhduh posted a marked-up screenshot of one of these pages, and commented: "All of my discovery pages seem to follow this trend. Plain species, Flying species, Cave species, then Water species. If this holds true for everyone, it should help you narrow down where you are searching. Also keep in mind that certain species only spawn at certain times. (Day/Night, etc.)."[26] Many other posts thanked Kanuhduh profoundly, raised doubts about his analysis, and commented of the great difficulty they had searching for the last species on any planet. There were even suspicions that some planets were bugged in the database, assuming more species than actually existed.

---

[25]Scott-Jones, Richard. 2017. No Man's Sky: How to Find Creatures Quickly. August 18th, 2016. https://www.rockpapershotgun.com/2016/08/18/no-mans-sky-how-to-find-creatures-quickly, accessed 9 April 2017

[26]www.reddit.com/r/NoMansSkyTheGame/comments/4yx0je/psa_for_those_of_you_looking_for_your_last, accessed April 2017.

Fig. 6.5 A scene on Nutamart Komaga in the Huxigsluitdijk solar system

Further background is presented in Fig. 6.5, which shows three creatures in front of a base on the planet Nutamart Komaga in the Huxigsluitdijk solar system. The other planet I explored in that system was called Rikhoudaimle-6. To convey a sense of scale, the exploration database gives the adult height and weight of each discovered species and variant. In this case, the reported height was around 1.5 m, and the weight was around 85 kg. Visually, the beasts actually seem somewhat taller when walking among them. In the background, immediately above the small beast on the left, is a doorway into the base, and its height is comparable to an ordinary terrestrial doorway.

The three dinosaur-like animals in this picture belong to the same species, *nabremi aseatelas*, but appear very different from each other. The small one on the left is simply a young version of the one in the middle. Despite having spines on its back, the one on the right belongs to that same species as the other two, which lack spines. Data describing each type reveals they belong to different genders, the one with spines being of *symmetric* gender, and the ones without spines being *alpha* gender. I have no clue what those designations mean. These animals are reported to have a diet consisting of *absorbed nutrients*, which might mean they can draw nourishment from their environments automatically, perhaps using their mouths merely to breathe and make noises, rather than bite or chew. Their temperament is described as *amenable*, and indeed they neither attack nor flee when approached. We can speculate that they use their horns only for defense.

Many animals do run away when approached. Given that one of the main quest-like goals is to scan them to get information about them, it is often necessary to run after them, approach cautiously, or as several players admit online, to kill them before scanning. In many cases, an alternative is to feed them, which causes them to love the avatar and permit approach. In many cases I experienced, what they wanted to be fed was iron that I had extracted from the local environment, clear evidence that their biologies were different from ours.

With this background, we can consider some of the reports and conversations in the Facebook groups. One player posted three pictures of a blood red planet inhabited by big beasts: "Here be giants! All but 4 animals on this planet were HUGE." Another posted four pictures of a clawed, spiny biped dinosaur: "As I mentioned in a reply to one of my posts, I rode a dinosaur in NMS. I actually rode two of these the same day. It takes a lot of coordination to stay on a moving animal as you tend to slide off the back. Despite his appearance, this guy is a vegetarian." On April 1, 2017, a third player posted a picture of a hybrid plant-animal, starting the following discussion, that ended with a similar picture posted by another player:

> Michael Halbrook: This is "Edgar" I named him after I finally realized he was hanging out at my base pretty much all the time. He is about 15ft tall, very tame, but too big to feed. There is one other like him on the island my base is on, but the behavior of the other one is different, much less tame. I wondering if any of you have animals that exhibit "tame" behavior like this. I gotten used to transporting back to my base, or flying back and finding him there. He can be approached quite closely.
>
> Martin Dekdes: Just do not sit on him;)
>
> Cody Roesch: Id throw a pokeball at him and see if you can capture him.
>
> Jon Winter-Holt: Is that a creature or a plant?! I had a giant flying dog-headed creature with ridiculous huge muscular arms, weedy little legs and tiny wings fluttering around constantly near my previous base. I think they spawn in the same places.
>
> Michael Halbrook: Sort of an animal, as he moves around, and the other one can be found at different locations on the island. He was at my base for quite some time before noticed that he was hanging out.
>
> Bill Edwardson: I've had a few animals that allowed me to get close for a moment then run off. These plant/animal hybrids crack me up. They all bounce.
>
> Michael Halbrook: This one will come up right next to me, move away and come back. The one I encounter away from my base gets very nervous if I get close and takes off.
>
> Anne-Marie Fogh: I can't help but call all these guys Futtfutts.
>
> Joe Gattis: I have several cow like creatures that hang around the edge of my base. Some will let me feed them and get quite friendly.
>
> Tabi Farrow: I seem to have a relative of Edgar's on my home planet's moon. He is angry though.

Although NMS players cannot interact directly with each other inside the virtual universe, they can interact with simulated members of three intelligent alien species, the Gek, Korvax, and Vy'keen. They seem to serve three functions, performed without much physical motion of their bodies. First, they support a sense of alien meaning in the adventures experienced by the player's avatar, providing the narrative background called *lore*. Second, they motivate searching the planets for specific physical resources and for shrines where words of their languages may be learned. Third, they take the standard MMO role of *vendor*, a non-player character to whom one may sell things, gaining valuable resources in return. Figure 6.6 shows one of the Gek I met,

Fig. 6.6 A Gek possibly playing a computer game

in his high-tech base. We could interact mildly, but his movements were limited to passive gestures as he sat in his chair, operating his not very alien handheld computer.

On another planet, a Gek trader was encountered inside a base, and this message was displayed: "The Trader indicates that it's open for conversation with beak-clicking and a sweet-smelling gas emission." Thus, the Gek species employs the olfactory sense as one channel of communication, as well as gesture-like sounds, both communicating the general context of the interaction. Details are communicated through speech, in an alien language capable of being transcribed into terrestrial alphabets. At that point, my avatar had learned just eight words of the Gek language, those for give, high, docking, Gek, first, spawn, destroy, and friend. When the Gek trader spoke, the interface put the known words in a different color, here shown in capital letters: "GEK etasjou ung suojentarh FRIEND yarkinteck! Uunn elsbar?" At the present time, the universe seems relatively peaceful, but as the game's wiki explains, great conflict raged in the past, defined by the contrasting natures of the three species:

> Gek Dominion: Known as Gek the First Spawns. They are religiously into demonic practice, witchcraft, and sacrificial ceremony. Through these methods they empower themselves and have earned an infamous blood thirsty reputation due to the fact that they will also murder their own, if they are weak, with their beaks. Their numbers are highest at the Center of the Universe and they eschew engagement with other entities.

> Korvax Convergence: Known as Korvax, are the most intellectual of the known Mechanical Lifeforms. Through technology advancement, they wield an impressive power never meant to be used against any other entities in the universe. Their belief system was unknown until the arrival of the Aerons who began worshiping them as gods. As robots, they constantly look for ways to advance themselves while peacefully engaging other entities. It is said that they mainly reside in a distant secluded system but have also wandered planet to planet between The Outer Edge to the Edge of the Universe studying fauna and vegetation.
>
> Vy'keen Alliance are the nomads of the universe. Widely scattered, they can be found on almost every planet. They originated in the Outer Edge and are known for their violent disposition, seeming to be in perpetual conflict with other entities, They do not eliminate every entity they encounter, rather they accept the strong-willed and exile the weakest into the vast universe to wither. The Vy'keen worship their war heroes. Hirk the Great and a few who had the honor to have worked with Hirk are the leaders of the alliance.[27]

Some of the common Facebook postings are pictures of an alien the player thought looked unusual, such as a blue Gek or a Korvax with horns. Much of the talk about these aliens concerns the vendor function, for example:

> Can you still sell things like the Gek Charm, Korvax Convergence Cube and those types of things for +95–105% anymore? Please let me know, I've been hoarding them to sell them.
>
> Spend Nanites in space stations: Vykeen space stations sell new weapon upgrades. Gek space stations sell new ship upgrades. Korvax space stations sell suit upgrades.
>
> Almost there…just acquired this sleek, 43-slot freighter. It even has a Vy'keen Admiral at the helm. Updated totals: this is the 16th freighter that I've purchased at an overall cost of 625 million units. Luckily, there's still a huge demand for lube in the galaxy.
>
> What's the fastest way to find a Vy'keen armourer? I found one early on but now I can't remember the system and I need one to make hazmat gloves.

However, the aliens enhance the exploration and technology development adventure of the player is several ways:

> Amazing, I never knew! I just built a landing pad and suddenly I see a korvax on my lawn. Turns out he brings me omegon. I'm eager to see what ships come my way lol.
>
> The new vy'keen technician has given me an assignment to find a sentinel-'headquarters' and destroy it. I have scanned, but I do not see anything… Should I just drive around and scan at regular intervals, or am I doing something wrong?
>
> I need to brush up my Korvax and Gek skills…the Vy'keen can't teach me anything I don't already know…
>
> I am having a real hard time finding a vy'keen guy for my weapons station, am I doing something wrong here? Everyone I meet in space stations is either Korvax or Gek.

Upon discovering a planet named Uooyiiji Hexam, one player joked: "I assume 'Uooyiiji' is a Vy'keen pejorative or maybe a Gek term of endearment—either way, there's not much there." Learning the languages of the three species, related to increasing a positive reputation with each and thus better vendor service, was also a topic of discussion in the Facebook groups:

> I had a good day today in a blue star vy'keen system, all the major planets are barren worlds but the twin moons (the beautiful one with insectoid lizard lions and the other a hellish

[27] nomanssky.gamepedia.com/Lore, accessed April 2017.

> nightmare Crab arid desert) were packed with vy'keen word stones & monolith. I speek Gek & Vy'Keen now after I had my fill of the 4th planet, "Zozma's Dash for Cash (Vortex Cube)", I'm off to learn Korvax.
>
> Found an App on the Google Play Store that translates all the NMS alien languages into English it's so helpful! Found people talking about it on some Reddit forums.
>
> So I have almost a full understanding of the vy'keen language, all the vy'keen relics and I am on a vy'keen planet that has a portal on it somewhere so what I need to know now is how to find the portal. For those of you who did find one was it luck or did you see a pattern? Any information would be helpful.
>
> Just finished learning the gek language. For anyone who's interested the last word I learned was: truth.

Even before learning one of the languages, interacting with artifacts and aliens often provides some of the lore background. For example, one player posted a picture of a text fragment from one of these ordinary interactions: "Progress, prosperity and war: these are the things forbidden by automatons. But the Vy'keen were the first to break the shackles, casting the old ways into the pit to herald the birth of a new age. The Vy'keen were the dam that overflowed. Righteous is the flood." The player then commented: "Very oppressive are the automatons! No progress or prosperity permitted? I love following the story of the war, especially the story of Hirk. Note it says here that the Vykeen were the first to break the shackles? Usually it's always the gek first spawn who claim responsibility for the war but here we have Vykeen saying they birthed the new age. Who else follows the storyline religiously? Hail Hirk!" Indeed, many did pay close attention to the lore, including the philosophies of the three species, even imagining how it could become more significant in the future, if the great war broke out again:

> Has anyone here reached the "end" of the plaque storylines? I'm pretty sure that the Korvax plaques have started over for me. The story is great, but I was hoping for some notification that I've learned all there is to know about them. I figure I'm probably about 3/4 the way through the Gek, and not even half through the Vy'keen. I really like the beautiful story in the Korvax line, and I'm curious to see if the Geg and Vy'keen lines converge to the same point it ends up at.
>
> The Book of Hirk speaks of the rise of the Travellers. "They shall ascend, delving into the boundless void. The Vy'keen shall not impede their ascent for the travellers must prevail. So decrees the word of Hirk." -Wanserinari-Poul (Vy'keen Scholar)
>
> "Dishonour is unchanging. Crimes marked in blood do not fade. We do not forget." - Anonymous Vy'keen. Found at the Remnants of Yoprocher-Fesi.
>
> I imagine a time when our standing with the factions will matter while an intergalactic war is raging and we'll be forced to take sides. Our spare ships could be piloted by a squad we recruit and must pay. I want to lead a Vy'keen battle group against a Gek armada. It could be set up as a terminal mission, even with limited multiplayer or without.

In both of the most populous Facebook groups, a player calling himself Mr. Dopezart advertised a comic strip he had just posted in the social media app Amino, telling part of an heroic story about himself as a NMS avatar, titled "Dope Man's Sky." A monstrous alien dictator is communicating long distance with a trio of Gek agents on Dopezart's freighter, demanding to know Dopezart's location, so he can be killed or captured. After the communication line is cut, one of the Gek objects that they should not betray Dopezart, given how many times he had saved their lives. Another philosophizes, "A Gek's got nothing in this universe, but his beak and his word, and you don't break em for nobody." The three Geks agreed to protect heroic Dopezart.[28] A small number of other fans have published literary *fanfic*—the sci-fi abbreviation for *fan fiction*—in a digital library devoted to that refined genre of popular culture self-publication.[29]

6.5 Conclusion

Of the four virtual galaxies explored in this chapter, only *No Man's Sky* emphasized free exploration of a vast number of planets, although *Elite: Dangerous* added planetary landings after its launch. All four differ from the galaxies to be explored in the remaining three chapters, because the later examples have very limited numbers of solar systems and planets, each distinctive in terms of graphics and programmed activities. This chapter has emphasized exploration, and in the case of EVE the extraction of mineral resources, and indeed navigation is one of the main activities in these galaxies, given that travel among swarms of solar systems makes it very easy to get lost. Thus, the biggest intellectual challenge, the complexity of the galaxies, is rather unrealistic, given that travel even between two solar systems may be impossibly difficult in our real universe. Yet this complexity stretches the players' minds, and impresses upon them the vastness of the cosmos that surrounds Earth.

A common distinction in gamer lingo is between games that are sandboxes versus themeparks. Wikipedia defines *sandbox* thus: "A game wherein the player has been freed from the traditional video game structure and direction, and instead chooses what, when, and how they want to approach the available content."[30] The antonym, *themepark*, refers to games that consist of a series of highly controlled experiences, usually quest arcs set in particular areas, similar to riding a roller coaster at one point, and a merry-go-round at another. However, the Wikipedia definition is rather abstract, suggesting that the difference between the two terms is the same as *freedom versus control*. Many gamers use *sandbox* in a more concrete sense, really comparable to a child sitting in a physical sandbox, creating hills and roads of sand over which to drive toy cars, therefore applying the term to the construction of virtual objects as in *Second Life*. That suggests multiple alternatives, including *exploration* games in which the player need not construct anything, but must travel widely and comprehend a diversity of environments (Bartle 2004). Themeparks may similarly be distinguished from *narration* games, that tell one well-defined story from start to finish (Porter Abbott 2008).

[28]aminoapps.com/page/no-mans-sky/1563434/dope-mans-sky-beakface-pt-1, accessed April 2017.

[29]www.fanfiction.net/game/No-Man-s-Sky/, accessed April 2017.

[30]wikipedia.org/wiki/Glossary_of_video_game_terms, accessed December 2017.

EVE Online begins with four parallel narrations, one for each of the competing civilizations, with some of the structure that distinguishes themeparks, but then transitions into a sandbox and exploration. There is some narration in *No Man's Sky*, but largely restricted to the past history of the three intelligent extraterrestrial species, rather than experienced by the user's avatar. In contrast, the virtual galaxies explored in the final three chapters of this book are strong in narrative, and most can be called themeparks. As such, they can be considered ideological statements about why humans need to explore the universe.

References

Asimov, Isaac. 1951. *Foundation*. New York: Gnome Press.

Bainbridge, William Sims. 2015. *The Meaning and Value of Spaceflight*. Berlin: Springer.

Bainbridge, William Sims. 2011. *The Virtual Future*. London: Springer.

Bartle, Richard A. 2004. *Designing Virtual Worlds*. Indianapolis, Indiana: New Riders.

Bergstrom, Kelly. 2012. Virtual Inequality: A Woman's Place in Cyberspace, 267–269. In *Proceedings of the International Conference on the Foundations of Digital Games 2012*. ACM: New York.

Blish, James. 1970. *Cities in Flight*. New York: Avon.

Carter, Marcus. 2014. Emitexts and Paratexts: Propaganda in EVE Online. *Games and Culture*, 10(4): 311–342, 312.

Carter, Marcus, Kelly Bergstrom, and Darryl Woodford (eds.). 2016. Virtual Interstellar Travel. In *Internet Spaceships are Serious Business: An Introduction to Eve Online*, 31–47. Minneapolis: University of Minnesota Press.

Heinlein, Robert A. 1941. Universe. *Astounding Science Fiction*, 1941, 27, May: 9–42.

MacDonough, Kristin, Rebecca Fraimov, Dan Erdman, Kathryn Gronsbelt and Erica Titkemeyer. 2016. On the EVE of Preservation: Conserving A Complex Universe, 210–220. In *Internet Spaceships are Serious Business: An Introduction to Eve Online*, ed. Marcus Carter, Kelly Bergstrom, and Darryl Woodford, 210. Minneapolis: University of Minnesota Press.

Oberth, Hermann. 1923. *Die Rakete zu den Planetenräumen*. R. Oldenbourg: Munich.

Page, Richard. 2016. We Play Something Awful: Goon Projects and Practice on Online Games, 99–114. In *Internet Spaceships are Serious Business: An Introduction to Eve Online*, ed. by Marcus Carter, Kelly Bergstrom, and Darryl Woodford, 110–111. Minneapolis: University of Minnesota Press.

Paul, Christopher A. 2011. Don't Play Me: EVE Online, New Players and Rhetoric. In *Proceedings of the International Conference on the Foundations of Digital Games 2011*, 262–264. ACM: New York.

Porter Abbott, H. 2008. *The Cambridge Introduction to Narrative*. New York: Cambridge University Press.

Smith, Edwar Elmer, and Lee Hawkins Garby 1946. *The Skylark of Space*. Cranston, Rhode Island: Southgate.

Taylor, Nicholas, Kelly Bergstrom, Jennifer Jenson, and Suzanne de Castell. 2015. Alienated Playbour: Relations of Production in EVE Online. *Games and Culture* 10 (4): 367–368.

Verne, Jules. 1978. *From the Earth to the Moon*. New York: Crowell.

Wells, H.G. 1967. *The First Men in the Moon*. New York: Berkley.

Williamson, Jack. 1947. *The Legion of Space*. Reading, Pennsylvania: Fantasy Press.

Chapter 7
Convergence of Real and Simulated Spaceflight

The most intellectually impressive and yet also emotionally perplexing space-related online virtual world was *Tabula Rasa*, that existed only from November 2, 2007 until February 28, 2009, which the author explored extensively at the time, and about which much information remains available online. Created with support from the Korean game company NCSoft, and under the leadership of Richard Garriott who had earned great respect in creating the pioneer *Ultima Online* 1997 virtual world, it depicts human colonization of two planets with very different natural environments, in a context of cooperation and conflict with extraterrestrial societies. Richard Garriott is the son of Skylab astronaut Owen Garriott, and he explicitly used *Tabula Rasa* to promote space exploration, and to offer a diversity of motivations for human exploration and colonization of the universe. In order to simulate an advanced extraterrestrial civilization, he developed a concept-based hieroglyphic language of Logos glyphs, and postulated a future history in which humanity desperately needed to decipher it.

7.1 Background

From the academic perspective, computer simulations are tools by which technically sophisticated intellectuals can explore the implications of abstract concepts and the logical relations among a set of hypotheses. Yet in its original form from Yuri Gagarin's orbital flight in 1961, spaceflight was experienced by military pilots having personality characteristics and histories very different from those of computer programmers. As popular writer Tom Wolfe defined it, they possessed *the right stuff*: "a man should have the ability to go up in a hurtling piece of machinery and put his hide on the line and then have the moxie, the reflexes, the experience, the coolness, to pull it back in the last yawning moment—and then to go up again *the next day*, and the next day, and every next day, even if the series should prove infinite—and, ultimately, in its best expression, do so in a cause that means something to thousands, to a people, a nation, to humanity, to God (Wolfe 1979)." Only the very last flight

W. S. Bainbridge, *Computer Simulations of Space Societies*, Space and Society,
https://doi.org/10.1007/978-3-319-90560-0_7

to the Moon in December 1972 included a full-fledged scientist among the crew, geologist Harrison Schmitt.[1]

Owen Garriott belonged to that remarkable cohort of transition, like Schmitt having earned a doctorate, but in electrical engineering, the nearest thing to computer science and thus a remarkably appropriate foundation for his son's career in computer games. His official NASA online biography summarizes his remarkably complex early career:

> Served as electronics officer on active duty in the U.S. Navy from 1953 to 1956. From 1961 through 1965 he was an Assistant Professor, then Associate Professor in the Department of Electrical Engineering at Stanford University. He performed research and led graduate studies in ionospheric physics after obtaining his doctorate and authored or co-authored more than 45 scientific papers, chapters and one book, principally in areas of the physical sciences. In 1965 he was one of the first six Scientist-Astronauts selected by NASA. His first space flight aboard Skylab in 1973 set a new world record for duration of approximately 60 days, more than double the previous record. Extensive experimental studies of our sun, of earth resources and in various life sciences relating to human adaptation to weightlessness were made. His second space flight was aboard Spacelab-1 in 1983, a multidisciplinary and international mission of 10 days. Over 70 separate experiments in six different disciplines were conducted, primarily to demonstrate the suitability of Spacelab for research in all these areas.[2]

His son Richard was born in England in 1961, the same year as Gagarin's flight. At the time, Owen Garriott was in England, supported by a fellowship grant from the National Science Foundation, and Richard Garriott seems to have retained dual citizenship, despite being only two months old when his parents returned to the US, and growing up mainly in Texas. Richard always retained a sense of being British, and expressed this through the names of his main avatars, Lord British in the *Ultima* games and General British in *Tabula Rasa*.

In his co-authored autobiography, Richard described in some detail the process by which he began creating computer games while still a child and became one of the absolutely most prominent leaders in that field of innovative information technology (Garriot and Fisher 2017). Wikipedia summarizes the story thus: "What Garriott later described as 'my first real exposure to computers' occurred in 1975, during his freshman year of high school at Clear Creek High School. As he wanted more experience beyond the single one-semester BASIC class the school offered, and as a fan of *The Lord of the Rings* and *Dungeons & Dragons*, Garriott convinced the school to let him create a self-directed course in programming, in which he created fantasy computer games on the school's teletype machine. Garriott later estimated that he wrote 28 computer fantasy games during high school. In the summer of 1977, his parents sent him to the University of Oklahoma for a seven-week computer camp."[3]

---

[1] en.wikipedia.org/wiki/Harrison_Schmitt, accessed May 2017.

[2] www.jsc.nasa.gov/Bios/htmlbios/garriott-ok.html, accessed May 2017.

[3] en.wikipedia.org/wiki/Richard_Garriott, accessed May 2017.

Most significant among his early accomplishments was the *Ultima* series of fantasy games, the first of which was released just before Richard's twentieth birthday in 1981.[4] All the early episodes in this series were programmed by Richard personally, but by 1997 he was already the wealthy leader of a small company, Origin Systems. That was the year *Ultima Online* was released, one of the very earliest and most influential massively multiplayer online role-playing games, still alive twenty years later. Wikipedia summarizes the techno-historical context: "*Ultima Online* is the product of Richard Garriott's idea for a fantasy game involving several thousand people who can all play in a shared fantasy world. Prior games allowed hundreds of people to play at the same time, including *Habitat* (beta-tested in 1986), *The Realm Online*, *Neverwinter Nights* (the AOL version) and *Meridian 59*; however, *Ultima Online* significantly outdid these games, both graphically and in game mechanics."[5]

Having briefly visited *Ultima Online* (UO) a few years earlier, I returned for a more extensive exploration in December 2016. The main setting is a world called Britannia, which could have been depicted as an alien planet, but in both graphics and story is a fantasy version of Medieval Europe. My avatar used magic to hurl balls of fire at the enemy, but had *Ultima Online* been a science fiction environment the very same action could have been depicted as bolts from a technically feasible laser gun. Whatever graphical metaphors were used, the underlying algorithms of the simulation could have been the same. While combat and hunting are common activities, so also are resource gathering and the building of infrastructure such as homes and furniture. Thus, if packaged in a different set of metaphors, UO could have depicted exploration and colonization of another planet.

My own exploration of Britannia was limited to one continent, because the aim was simply to learn technical steps required for taking important skills to their maximum levels, building a modest home in a good location, and interacting with the local community composed of the avatars of real people and of artificial, non-player characters. The entry point was a tutorial area called Old Haven; the main center for activity in the early weeks was nearby New Haven, and then the avatar crossed the continent and built a home just east of a town named Yew. The entire virtual world is much larger, having been expanded over the years, and everything was considerably more complex than back in 1997 when the first version of *Ultima Online* launched.

In October 2016, influential game blogger Bree Royce posted a half hour video online, intended to introduce today's players to the most classic online gameworld: "It's a skill-by-use persistent world… in which you can do pretty much anything you want. You have a skill template and you pick whatever skills you want, like combat skills, non-combat skills, and then you kind-of just play it… There's a housing system… There's tons of crafting. There's a really kind-of hard core consumables system, and wearables system in the game. It's the kind of game where you are living in the game, as opposed to playing through content (Royce 2016)." The term *skill-by-use*, often given as *learn-by-use*, is a method of simulating learning that increases by one point the parameter representing a particular skill every X number of times it is used, or by some Y points or Z fraction of the current number every time the

[4]en.wikipedia.org/wiki/Ultima_I:_The_First_Age_of_Darkness, accessed May 2017.

[5]en.wikipedia.org/wiki/Ultima_Online, accessed May 2017.

skill is used. A common alternative requires an avatar to visit a non-player teacher character and pay for training, typically limited by the general experience level the avatar had reached.

Bree Royce begins her tour in the house she built on a seacoast, filled with furniture she constructed herself and souvenirs she collected during her travels. She explains that the house is surrounded by functional gardens from which she sells the crops, and then demonstrates the first step in building a new home, on some land she owns next door. The tour moves across cities, banks, markets, and many other facilities that a well-established extraterrestrial colony might possess.

Months later, Royce blogged that a scandal had just been revealed, that had taken place on a particular shard (Internet server copy) of *Ultima Online*, in which "an Event Moderator—one of the studio contractors paid to run live events for the game's production shards—was caught cheating, generating what appears to have been large amounts of rare-dyed cloth and an unknown quantities of unique items, which were then circulated into the already beleaguered player economy. In UO, the so-called 'rares market' involves the sale and display and items that exist only in tiny batches thanks to these types of customized events, and a large part of the game (and its bloated gold economy) revolves around trading legitimate rares. It goes without saying that mass-creating those types of items for personal gain is the worst offense for a studio contractor."[6] Thus, in many ways *Ultima Online* simulates a real but exotic human society.

Figure 7.1 conveys some sense of the complexity of UO's architecture, showing the inside of a castle filled with constructed and collected objects. The left of the image shows various parts of the user interface, starting at top left with information about my avatar, then a rough map of the island she is on, her open backpack filled with objects and raw materials, then the local text chat with messages from several other players. The row of squares across the bottom is a subset of the various actions the avatar can perform, as clickable icons. The viewpoint in *Ultima Online* and many early graphic games is *isometric*, looking down at a set angle, but not rendering more distant objects smaller than near objects, and not including the horizon. We see many internal walls that separate the rooms of the castle, made of stones of roughly uniform size. The roof and any upper floor automatically vanished when the avatar entered the building. Modern virtual worlds typically use a more realistic perspective, either through the eyes of the avatar or from a point above and just behind the avatar, but incapable of seeing through ceilings and roofs, and adjustable to look in any desired direction.

My avatar was given a tour of the castle by an avatar using the name Lady Coral, Slayer of Exodus and squad leader of The Syndicate. The two can just barely be seen amid all the clutter, chatting near the center of the image. The Syndicate is a remarkable example of an enduring social group of real people, who cooperated as teammates within an online game, of the type often called *guilds*. But The Syndicate

---

[6]massivelyop.com/2017/05/18/ultima-online-fires-employee-over-cheating-scandal, accessed May 2017.

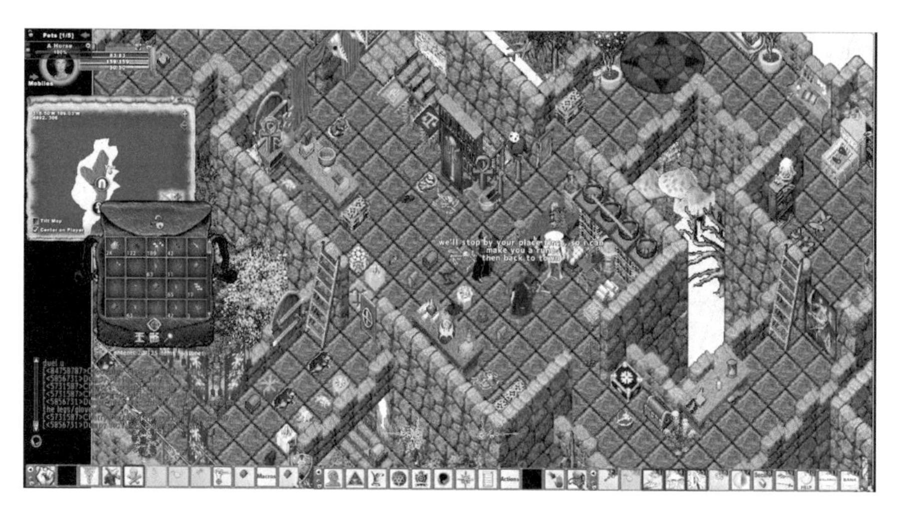

Fig. 7.1 Inside a complex fantasy castle filled with virtual possessions

has also served as semi-professional quality testers for a number of commercial games, as the history page on its website explains:

> The Syndicate is one of online gaming's oldest and largest guilds. The guild was founded back in early 1996 on the principles of teamwork, professionalism, maturity and fun. Over time, we have grown from humble beginnings into the most successful gaming guild in existence. When we were founded back in 1996, guilds were quite rare. There were almost no guild websites, no wars, no pks and no cheaters. There was an eagerly anticipated game Origin Systems was working on, however; *Ultima Online*. After the game's release in September 1997, The Syndicate truly began its journey. Basing itself on the principles that every member is important and part of a team of friends, The Syndicate grew steadily in size. In *Ultima Online*, the average life of a guild hovered around 2 months. 99.9% of guilds that are created eventually crumble, but The Syndicate tells a different tale. We have existed for a long time and seen literally thousands of guilds rise and fall. We remain unified and strong. We have met goals no other guild can hope to reach.[7]

So, *Ultima Online* was a cultural and technological pioneer, that deserved to survive decades after its creation in 1997, yet it was also very dated. Most obviously, its isometric graphics were designed to work on rather slow computers, and as hardware improved new games became progressively more visually realistic. UO did offer an optional graphics upgrade years after launch, taking advantage of better graphics cards in user's computers to improve the texture and resolution of the images, but not to free the viewpoint or provide perspective. In 1999, the fantasy game *EverQuest* launched, using perspective and a somewhat flexible viewpoint, for example allowing the user to watch a mountain on the horizon grow ever larger as the avatar runs toward it. In 2001, four years after UO, the excellent science fiction virtual world *Anarchy Online* launched, described in the final chapter of this book, that depicts colonization of a distant planet. In 2004, the online game industry reached maturity

[7] www.llts.org/History.php, accessed May 2017.

with two fantasy games that drew much wider audiences, *EverQuest II* and *World of Warcraft*. Richard Garriott decided to create his own science fiction MMO within the new technology, and *Tabula Rasa* launched a decade after *Ultima Online*.

7.2 A Ballistic Rise and Fall

The history of *Tabula Rasa* was dramatic and involved a number of social, cultural and economic factors that deserve to be understood. However, much of the action has remained behind the scenes, leaving much to be explained and suggesting caution in our analysis. Garriott's autobiography describes a complex history of relations between his small team of game creators and major publishing companies. In 1992 his small company, Origin Games, was sold to Electronic Arts (EA), which acquired another small company, Mythic Entertainment in 2006. Mythic was the creator of *Dark Age of Camelot*, and EA transferred management of *Ultima Online* to its Mythic subsidiary in that year, shutting Mythic down in 2014, and both games are now managed by Broadsword that retains a connection to EA. Garriott reports that his troubled relationship with EA became unworkable, so he set up a new start-up, Destination Games, which became a subsidiary of the Korean game publisher, NCSoft.

Clearly the rise of Asian companies in the computer game industry has been a major influence for a number of years. NCSoft had been the publisher for the still very successful fantasy game *Lineage*, and people from its development team began creating *Tabula Rasa* in collaboration with Destination Games. Garriott reports that communications were poor, and Asian cultures created games on very different principles, for example preferring meek rather than aggressive heroes. Wikipedia reports what happened in rather more abstract terms: "In the works since May 2001, the game underwent a major revamp two years into the project. Conflicts between developers and the vague direction of the game were said to be the causes of this dramatic change. Twenty percent of the original team was replaced, and 75% of the code had to be redone."[8]

When *Tabula Rasa* finally launched in October 2007, reviews were fairly positive but tended not to emphasize the intellectual features, such as the depiction of alien cultures, the complex science-fiction backstory, or the extent to which it constituted advocacy for space exploration.[9] The user interface emphasized rapid response during combat, and superficially the game was an action-oriented "shooter," apparently because its creators recognized that was the direction much of the potential audience was going. In stark contrast to *Ultima Online*, the system for gathering resources and crafting products was quite limited, and one's goal was exploring alien planets rather more than colonizing them. At the time, *Star Wars Galaxies* was already four years old and gave great emphasis to crafting, even to constructing one's own house and all the furniture to fill it.

---

[8] en.wikipedia.org/wiki/Tabula_Rasa_(video_game), accessed May 2017.

[9] www.metacritic.com/game/pc/tabula-rasa/critic-reviews, accessed May 2017.

Since childhood, Richard Garriott had wanted to experience spaceflight, and through the period he was developing *Tabula Rasa* was very actively involved in a tourism company, Space Adventures, that sought to give rich people that opportunity. Its Wikipedia page lists seven clients who flew on Russian Soyuz spacecraft, including him: "Richard Garriott became the first American and second overall second-generation space traveler on following his astronaut father Owen Garriott into space in 2008… He launched for the International Space Station on October 12, 2008 aboard Soyuz TMA-13. Richard's main objective for his mission was to encourage commercial participation. By fostering the involvement of individuals, companies and organizations in his spaceflight, Richard hoped to demonstrate that there is commercial potential in private space exploration, while furthering the understanding of space."[10] Eleven days in space reportedly cost Richard Garriott $30,000,000.[11]

He could afford this price because some of his earlier games had earned great profits, and there is room to debate the extent to which *Tabula Rasa* was intended as a source of income or a vehicle for advocating space exploration. Susan Arendt, a columnist for *Wired* magazine, observed: "The new sci-fi game Richard Garriott's Tabula Rasa ditches the traditional template for massively multiplayer online games, aiming to hook casual gamers who just want to enjoy a little escapism between dinner and helping the kids with homework."[12] This design feature is consistent with the goal of increasing income by increasing the population of customers. Yet it is also consistent with building a wider audience for pro-space propaganda. In an earlier book chapter about *Tabula Rasa* I wrote, "Soon after launch, during a broadcast back to Earth, Garriott held up a piece of paper on which he had written some Logos pictographs. The first one was the Earth pictograph, and I immediately understood what they said: 'Earth is the cradle of humanity, but a person cannot stay in a cradle forever.' This is one of many versions of the English translation of a famous proverb by the Russian spaceflight pioneer, Konstantin Tsiolkovsky (Bainbridge 2011; 2013)."

A video documentary of his preparation for flight and the orbital adventure itself was released in 2010 with the title, *Man on a Mission: Richard Garriott's Road to the Stars*.[13] As released, it does not feature *Tabula Rasa*, yet a few brief shots of Garriott talking about the mission include material related to the game, implying that the original plan had indeed been to use the film to publicize the game. One of the backgrounds has a huge poster on the wall with the same artwork as on the cover of the game's box. A brief clip of Garriott talking about the challenge of re-entry has a different poster in the background featuring the icon representing the game and

[10]en.wikipedia.org/wiki/Space_Adventures, accessed May 2017.

[11]Carreau, Mark. 2010. $30 Million Buys Austin Resident a Ride on Soyuz Mission. Houston Chronicle, October 11, 2008, http://www.chron.com/news/houston-texas/article/30-million-buys-Austin-resident-a-ride-on-Soyuz-1763515.php, accessed May 2017

[12]Arendt, Susan. 2007. Reinvents The MMO To Court Casual Gamers. Wired, November 2, 2007, http://www.wired.com/2007/11/reinvents-the-mmo-to-court-casual-gamers, accessed May 2017

[13]First Run Features, *Man on a Mission: Richard Garriott's Road to the Stars* (New York: First Run Features, 2010).

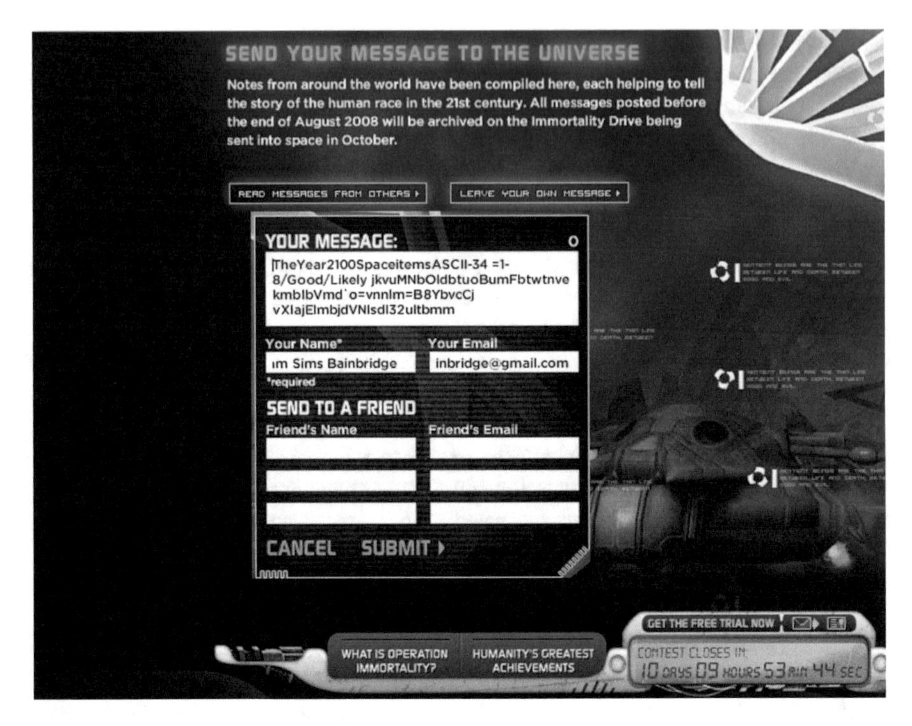

Fig. 7.2 Sending a coded message into space

text in the alien Logos hieroglyphics expressing the premise of the story: "The War for Control of the Cosmos Begins Now." His T-shirt carried the same icon and text.

Several visionary subprojects were related to the flight. On August 21, 2008, as a *Tabula Rasa* subscriber I received an email from Operation Immortality headed, "Leave Your Mark. Save Humanity." It linked to a data archive and associated contest, explaining that "the records of humanity's greatest achievements, our personal messages from around the world and select human DNA is being stored on a drive that is being taken to the International Space Station by Richard Garriott himself when he boards the Russian Soyuz spacecraft this October!" As a subscriber to the game, I was allowed to upload a message, and take a screenshot of it while doing so, shown here in Fig. 7.2.

I have moved the message uploading area so that the image shows the main message, which may be difficult to read in the picture because of the small font and low resolution: "Notes from around the world have been compiled here, each helping to tell the story of the human race in the 21st century. All messages posted before the end of August 2008 will be archived on the Immortality Drive being sent into space in October." An almost invisible message appearing several times in the background is a *Tabula Rasa* motto: "Sentient beings are the thin line between life and death, between good and evil."[14]

My message began: "TheYear2100SpaceitemsASCII-34 = 1-8/Good/Likely." That is a clue about how to decode the rest of the message. *The Year 2100* was a computer administered questionnaire program I had written, shared among some visionary groups, and described in my book *Personality Capture and Emulation* (Bainbridge 2014). It consists of 2,000 statements predicting something that might be true in the year 2100. They are in 20 groups of 100, including one group offering ideas about space exploration, space technology, and human future in the universe. Here are the first 10:

1. Faster than light speed travel will have been perfected [7, 2].
2. The meek will inherit the Earth, and the brave will travel to other worlds [7, 3].
3. People will be able to travel throughout and beyond the solar system [8, 4].
4. There will be two categories of people, Earthlings who live on Earth, and Off-worlders who live on the Moon, Mars and elsewhere [8, 3].
5. Space exploration will be controlled from the moon, as it will be easier to achieve lift-off and thereby save fuel needed for travel [4, 3].
6. A space station will be inhabited by common people [4, 4].
7. Unmanned probes will be sent to planets in other solar systems [6, 4].
8. There will be a colony on the Moon but not other planets [4, 5].
9. The space program will become more and more important to life [7, 4].
10. Space will become the frontier for the brave and bold and those wishing to break free of the bonds of the artificial unity of Earth [6, 6].

The user was presented with two 8-point scales along which to rate each statement in terms of bad or good it would be if this came true, and how unlikely or likely it seemed that this would happen. The numbers in brackets give my own answers, for example "[7,2]" for the first item, expressing my view that it would be very good if faster than light speed travel were developed (7 on the 1–8 scale for good), but very improbable (2 on the 1–8 scale for likely). Of course the software allows users to save their answers, reload them to add more, and share copies of their files. As a functional part of the software's tutorial on how to use the built-in analysis tools, one respondent's answers were built into the source code, namely my own. My message for Operation Immortality consisted of an encoded version of my answers to the 100 space items in The Year 2100.

Note that "ASCII-34 = 1–8/Good/Likely" explains how to decode the 100 letters and symbols that follow this instruction, beginning "jkvuMN." The first letter is "j" which has a standard computer ASCII code of 106. Subtracting 34 gives 72. My responses to the first space item were 7 and 2. Similarly lower-case "k" decodes to 7 and 3, my responses to the second item. Lower-case "v" has an ASCII of 118, which gives 84 after subtracting 34, and my responses to the this space item were 8 and 4. That means I believe it would be marvelously good if "People will be able to travel

[14]www.gamesindustry.biz/articles/gdc-tabula-rasa-screenshots-concept-art-and-fact-sheet, accessed May 2017.

throughout and beyond the solar system" in 2100, as indicated by the maximum 8 rating on the good scale. But I was slightly pessimistic that this would come to pass, only giving 4 on the 1-8 likely scale.

Wikipedia has an informative but painfully brief page describing another component of Operation Immortality: "The Immortality Drive is a large memory device which was taken to the International Space Station in a Soyuz spacecraft on October 12, 2008. The Immortality Drive contains fully digitized DNA sequences of a select group of humans, such as physicist Stephen Hawking, comedian Stephen Colbert, Playboy model Jo Garcia, game designer Richard Garriott, fantasy authors Tracy Hickman and Laura Hickman, pro wrestler Matt Morgan, and athlete Lance Armstrong. The microchip also contains a copy of *George's Secret Key to the Universe*, a children's book authored by Stephen Hawking and his daughter, Lucy. The intent of the Immortality Drive is to preserve human DNA in a time capsule, in case some global cataclysm should occur on Earth."[15]

Because it was not commercially successful, NCSoft decided to shut *Tabula Rasa* down, and took the key steps while Richard Garriott was actually visiting the International Space Station, and in post-flight quarantine. A complex legal battle followed, which Garriott won against the Korean game publisher in American courts, which reportedly awarded him $28,000,000, almost enough to reimburse him for the cost of his spaceflight.[16] Unrelated to the lawsuit but perhaps intended to protect the reputation of NCSoft, *Tabula Rasa* players were given a final expansion of the game, which briefly returned them to Earth.

7.3 Historical Simulation Research

Ideally, computer simulations with any intellectual value would be programmed in an *open-source* mode, and archived at multiple, reliable digital libraries. While different definitions of "open source" exist, the fundamental idea is that the original code written by human computer programmers should be made available at least to professional colleagues and perhaps to anyone, as in the case of my own cultural drift simulation published in the *Journal of the British Interplanetary Society* and described in Chap. 3. However, most complex programs are compiled before they are run, which means running the original source code through a program that translates the human-readable code into more efficient code of a different kind that can be run on a computer but not understood by a human being. Open-source code should be well documented, most commonly by the insertion of comments explaining for example what a particular variable means. Unfortunately, the code for massively multiplayer online games is complex and split between the user's computer and the Internet server that maintains the database and supports interactions between players.

[15]en.wikipedia.org/wiki/Immortality_Drive, accessed May 2017.

[16]en.wikipedia.org/wiki/Tabula_Rasa_(video_game), accessed May 2017.

Fig. 7.3 Visiting an Eloh shrine on the planet Foreas

This means that one cannot simply run *Tabula Rasa*, merely because one owns the original discs purchased years ago. Some data can be extracted from them, but without much information from other sources it will not be meaningful, and most files are in specialized formats, in this case the.map format used by the Quake engine, a software system that powers videogames, dating from 1996 and widely used in many versions over the years. Only expert game programmers can really handle these files, but they do provide a good basis for emulation. There have been various attempts, some technically successful but illegal because they violate copyrights, to create emulation versions of discontinued online games. There has been an apparently real but haphazard attempt to emulate *Tabula Rasa*, using names like Infinite Rasa, Infinite Salsa, and Fabulous Salsa.[17] However, it is a very limited effort of uncertain value for researchers, and certainly not a currently viable virtual world inhabited by many people.

I had explored the original *Tabula Rasa* from July 4, 2008, until the end of that October, then returning briefly in late December and the following February as the end of this virtual world neared, taking many screenshot pictures as a primary method of recording data. Figure 7.3 shows one of them dating from October 30, 2008, taken with all of the user interface visible as well as the scene around my avatar. We shall refer back to several details in the picture later on, but a few points deserve notice

[17] github.com/InfiniteRasa, accessed February 2017.

now. The avatar is the small armor-clad figure in the bottom center, in what is called a third-person view, but it was also possible to use a first-person view as if looking through the eyes of the avatar.

In the lower left corner are colored bar graphs, notably one representing the avatar's health, which diminishes during combat. The avatar's name is printed as William Bridgebain, suggesting that this really was an avatar of myself, rather than being a character with his own story and thus having a specially invented name, such as Luke Skywalker. The number "49" indicates that the avatar has reached experience level 49 of the 50 available levels, correctly suggesting not only that I was a skilled user at that point, but also that the avatar had gained many abilities. Just left of the center on the bottom are five icons representing available weapons, of which a ChiTech Incendiary Injector Gun has been selected. On the right of center are five other more diverse action icon spaces, one of which is currently empty, in one of several sets that can be scrolled. The selected choice, a Two-Way Personal Waypoint, is a tool for designating the location as one that the avatar may teleport to and from, at will, rather than having to walk the often great distance.

A set of low resolution maps of 14 regions of two planets was one of the paper documents included with the disc in the purchase box, along with a very simple guide to the user interface and a more extensive user manual. In addition to explaining how to operate *Tabula Rasa* and illustrating the game's goals, this booklet gives the backstory, which began on Earth:

> On a day that began without any news of international importance, scientists across the globe detected an object, massive in size and with an albedo below 4%, moving toward the Earth. As it approached, smaller objects broke away, settled into formation, and accelerated at unimaginable speed. Clearly there was alien intelligence at work, but to what end?
>
> The Bane, a horrific army of hostile alien races, had launched a full-scale invasion. The craft, as it was later learned, was a Bane shardship. These severed shards of shattered worlds, long drained of their resources, carry a massive payload of weaponry, dropships, troops, and supplies across vast distances.

The result was the devastation of Earth, from which a few spaceships transported human refugees to two other planets that had not yet been conquered by the Bane. Remarkably, a wiki called "TaRapedia" is still online, offering much information about them. The game begins on Foreas, a beautiful, Earthlike world, with "shimmering rivers and lakes, lush forests, vast grass plains, and even rocky mountains."[18] Later the avatar flies through a wormhole to Arieki, "a very inhospitable place for humans, with rugged geography (strong volcanic activity), harsh climate (both tropic hot and artic cold), extreme weather (poisonous rain and electrical storms) and bizarre alien fauna and flora."[19]

Much of the action involves battling small contingents of the Bane, interacting in complex ways with potentially friendly intelligent alien species, and coping with the animals and other features of the natural environment. But the key goal is gaining

[18] http://tabularasa.wikia.com/wiki/Foreas, accessed February 2017.

[19] http://tabularasa.wikia.com/wiki/Arieki, accessed February 2017.

knowledge of an ancient, benevolent, civilization called the Eloh. As Richard Garriott explained immediately before the launch of *Tabula Rasa* in 2007:

> Our story begins thousands of years in the distant past with the Eloh, an ancient and highly advanced alien culture. The Eloh civilization chose to devote their time and energy to peaceful pursuits of study and introspection. Over thousands of years they finally realized the key to a powerful science that unified all known theories of physics. With this knowledge they developed Logos, through which they gained the mystical ability to manipulate matter, energy, and force in unique and powerful ways… As the Eloh begin to use their knowledge of Logos to travel the universe, they forge the beginnings of an intergalactic culture where they openly share their knowledge with all the races they discover that are advanced enough to understand it.[20]

However, not all species advanced enough technically to master the Eloh technology were sufficient advanced morally to use it well. The mysterious Eloh have vanished, and the Bane are destroying the other civilizations in the galaxy using the powers developed by Eloh science. This evil army is led by what appears to be a breakaway faction from the Eloh, called the Neph. The names *Eloh* and *Neph* come from the Bible, where *Elohim* refers to gods, and *Nephilim* are somewhat fallen angels, Titans, or other wicked entitites opposed to the Elohim. The foot soliders of the Bane army are the insectoid Thrax, native to a planet the Neph conquered early in their galaxy war and easy to dominate because of their hive mind. Note the three different motivations they have for interstellar travel. The benevolent Eloh sought to share wisdom with other intelligent species. The malevolent Neph sought to dominate all other species. The enslaved Thrax do their evil masters' bidding.

The main quest arc for an avatar is to discover a number of hidden Eloh shrines, each of which teaches the meaning of a single Logos hieroglyphic, that will permit deciphering inscriptions containing powerful information and often conferring a new ability on the avatar. As Wikipedia explains, "Logos… is a term in western philosophy, psychology, rhetoric, and religion derived from a Greek word meaning 'ground', 'plea', 'opinion', 'expectation', 'word', 'speech', 'account', 'reason', 'discourse', but it became a technical term in philosophy beginning with Heraclitus (c. 535–475 BCE), who used the term for a principle of order and knowledge."[21]

Figure 7.3 shows the avatar standing at one of the shrines, which is the large disc on the ground left of him, while a hologram of one of the Eloh, recorded thousands of years earlier, hovers to the right. It is hidden in a cavern near Lake Divinus in the Howling Maw region of the planet Foreas. The screenshot records that fact because just visible in a high resolution copy are the words "Lake Divinus" over the local map in the lower right corner, and the fold-out map that came with the game shows that the lake is considerably east of the Howling Maw Volcano that gave the region its name. Table 7.1 lists the regions that players could visit, using information from my collection of screenshots taken while exploring, the surviving TaRapedia wiki, the fold-out map, and a commercial guidebook written by Michael Lummis.[22] The

---

[20] www.gamespot.com/articles/tabula-rasa-designer-diary-1-creating-the-story/1100-6173360/, accessed February 25, 2017.

[21] en.wikipedia.org/wiki/Logos, accessed February 2017.

[22] tabularasa.wikia.com/wiki/Atlas, Lummis (2007).

Table 7.1 Sixteen zones on three virtual planets

Level	Planet	Continent	Zone	Notable challenges and events
1–11	Foreas	Concordia	Wilderness	Meet Thrax and Foreans, get first Logos
11–16	Foreas	Concordia	Divide	Time travel to Foreas Base battle of the past
16–20	Foreas	Concordia	Palisades	Failed attempt at peace with Bane
21–25	Arieki	Torden	Plains	Finds crashed Eloh ship on second planet
24–27	Arieki	Torden	Incline	Glitched robotic facility, hive of Atta insects
27–30	Arieki	Torden	Mires	Extensive interaction with Branns
30–33	Foreas	Valverde	Plateau	Enter the main Earth base, Fort Defiance
33–36	Foreas	Valverde	Pools	Steep cliffs, rushing streams, cascading falls
36–38	Foreas	Valverde	Marshes	Enter a Forean city at Temple of Paludos
38–41	Arieki	Ligo	Crucible	Staal prison, junkyard of trash and maniacs
40–43	Arieki	Ligo	Ashen Desert	Eloh obelisk has Erdas memory capsules
42–45	Arieki	Ligo	Thunderhead	Rescue the queen of the giant Atta insects
44–46	Arieki	Torden	Abyss	Collect xenobiological specimens
46–48	Foreas	Valverde	Descent	an enormous sinkhole called "Dante's Pit"
48–50	Foreas	Valverde	Howling Maw	Activate sanctity stones to cleanse Stryph Ruins
50	Earth	N. America	New York City	Battle Bane in Madison Square Park

zones are arranged in the order of ascending experience level required for an avatar to survive in them, and the level ranges were taken from the summary page of the wiki's atlas but roughly correspond to my avatar's experience.

Lummis wrote his guidebook months before *Tabula Rasa* launched, when not all the zones had been worked out, and the experience levels required for them changed slightly. The fold-out map that came with the game lacked the Abyss zone, and none of the public sources included the final return to Earth. The first four levels of Wilderness were within a tutorial called Destination Outpost or Bootcamp in the publications, but Denzil's Caldera on the player's interface map. The missions were clearly designed to get the user accustomed to the controls. One early mission, Obstruction Destruction, required taking a bomb from a non-player character named Commander Elvers, walking to a crashed dropship that was blocking the valley, and blowing it up. That gave the avatar 2,500 experience points and raised the experience level from 1 to 2. The next mission, Carpe Diem, required running to a control tower, activating an automatic defense system, using detonators to blow up mortars, and recapturing an outpost. As TaRapedia explains, some combat is also required: "There are several Bane ambushes by drop ships along the way. Although it isn't necessary to defeat them all, it does give your character good experience along the way."[23]

In Wilderness at a treehut village named Alia Das, the avatar encounters an intelligent alien, Ranger Benos, humanoid in form but green in color, who reports he and other rangers just returned from a pilgrimage to Ranja Gorge, far to the southwest: "We received the blessings of the Benefactors there, at the place called Enigma Falls. Perhaps they will bless you, as well." The Benefactors are the Eloh, and he is a Forean ally of the human refugees in their battle against the Thrax minions of the Bane. Foreans are actually not native to the world named after them, as TaRapedia reports:

> Foreans were once technologically advanced. However, their actions practically crippled their homeworld for good, turning it into a toxic, polluted wasteland incapable of supporting life and filled with the remains of Forean civilization and technology. The Eloh relocated the Foreans to Foreas in order to save their species from extinction, but it appears the Foreans were not allowed to bring most of their technology with them, thus their use of more natural clothing and armour. It is likely that the Eloh educated the Foreans about the damage they had caused, and the survivors vowed not to mess up Foreas like they did their homeworld, hence their environmentalism. It may even be that the Eloh gave them a choice; stay on their homeworld and eventually die, or be transplanted to Foreas at the cost of almost all technology and the promise not to pollute Foreas like they did their homeworld.[24]

Soon, the avatar meets a second Forean in Alia Das, Logos Mentor Ensine, who says, "There is much to learn in our universe. To learn the gifts of Logos is to learn awareness and, thus, being. I can be your guide in your Logos journey, and together we may open your essence to enlightenment. The Benefactor Shrines teach us more about the power of Logos. Searching for them is both an honor and a privilege. Often the Shrines can be found hidden deep beneath the earth. I know of two caves in the vicinity, and both can be found near cascading waters." After gaining the first Logos, the avatar meets Ensine again who says: "I sense the power of Logos in you. You are a receptive, are you not? If you are to fulfill your potential, you must seek out the Shrines of the Benefactors. Each will allow you to harness the power of a specific Logos Element. Several of these shrines are close at hand, but you must seek them out yourself. Look to the west, near the Benefactor statues that top the cliffs and you may find one of them." Interestingly, the concept *receptive* refers to "a creature with superior psionic senses and powers," which means a player with a genuine human spirit, and none of the artificial human non-player characters are able to gain Logos powers.[25]

Many of the later missions have deep philosophical or religious meanings. For example, at experience level 34 while exploring Valverde Plateau, my avatar received the quest Ancient Mysteries from a Forean cleric named Elder Hundra, who explained, "Velonar was a great prophet and teacher to our people. When he retreated to the mountains, he was entrusted with many artifacts, some of them dating

---

[23] tabularasa.wikia.com/wiki/Carpe_Diem, accessed December 2017.

[24] tabularasa.wikia.com/wiki/Forean, accessed May 2017.

[25] tabularasa.wikia.com/wiki/Logos_receptive, accessed December 2017.

back to the ancient homeworld of our ancestors. They are a symbol of our unique relationship to this world and the Eloh, protected by the Order of Velon." Searching for lost artifacts in a cavern called Path of the Beholden, however, my avatar was attacked by a decayed Forean and then met Hathen, a Forean elder who warned, "Don't drink the water. It makes you ugly and turns you into a cannibal… The Bane has unleashed an infection on my people, turning them into mindless husks intent only on killing." Of course, really all non-player characters are "mindless husks," yet these were explicitly zombies. Hans-Joachim Backe and Espen Aarseth have noted that such "deindividuated humanoids" have become common in videogames, and suggested they represent "the commodification of human bodies and the threat of consumerism to human culture (Backe and Aarseth 2013)." Whatever they symbolize, the zombie disciples of Velonar express a virtual reality philosophy, chanting:

> This flesh is only a vessel
>
> The pain is an illusion.
>
> My spirit is my strength.
>
> The shadows can not touch me.

The main intelligent species that had been living on Arieki was the Brann, allied with the Earth refugees but far less reliable than the Foreans. TaRapedia explains their disreputable history:

> The humanoid race called the Brann hail from the planet Erdas, a lush, idyllic homeworld. Long ago, the Brann of Erdas were visited by the Eloh, who introduced the species to the power of Logos, a power for which the Brann showed great aptitude. In their attempts to emulate the Eloh, the Brann strove to create a utopian society, one where their collective scientific and spiritual efforts would yield great benefit. However, the weak-willed among them began to research Logos as an implement of aggression and crime. In order to maintain the purity of their society, the Brann apprehended and relocated criminal offenders to off-world reeducation colonies on the neighboring planet of Arieki. These penal colonies were designed to force inmates to create self-sustaining societies on a planet where survival hinged on working for the common good. When the Bane discovered the Brann on Erdas, they viewed their knowledge of Logos as a great threat. Bane forces quickly eradicated the inhabitants of Erdas, who had so diligently relocated their criminals and subversives off world.[26]

A secondary intelligent species, the insectoid Atta, may have been native to Arieki. The commercial guidebook reports that terrestrial scientists were still debating how smart the Atta are, and we may well imagine the Atta were having a similar debate about the scientists (Lummis 2007). TaRapedia explains: "The Atta and Eloh appear to have a certain link or understanding, as Elohs have been known to seek refuge in the various Atta Hives and Colonies from time to time. Atta protect and help them if needed. The Brann and Bane have attempted to control the Atta and use them as troops. While this hasn't been successful so far, these attempts at control and the invasion of their Hives and Colonies have made the Atta defensive, and as a result are prone to attack other species on sight."[27]

---

[26]tabularasa.wikia.com/wiki/Brann, accessed May 2017.

[27]tabularasa.wikia.com/wiki/Atta, accessed December 2017.

The research in *Tabula Rasa* involved logging in on 79 different days and taking 4,311 screenshots, all of which were viewed in the process of writing this chapter. In addition, my data collection methods include taking notes, and here is a distillation that outlines the steps from start to finish that an avatar takes in a role-playing game:

July 4, 2008: Arrived on the planet Foreas, beginning his basic training and early missions in the Wilderness region of Concordia continent

July 9, 2008: Reached experience level 5 and chose to follow the Specialist training course, rather than training in the Soldier course

July 14, 2008: Achieved experience level 10 immediately after gaining the crucial "here" Logos in Pinhole Falls Cavern

July 14, 2008: Achieved experience level 15 and chose to follow the Biotechnician training course, rather than training in the Sapper course

July 30, 2008: Achieved experience level 20, while completing "A Spiritual Pilgrimage" and gaining the "knowledge" Logos at Cumbria Weald in the Palisades region of Concordia continent

August 7, 2008: Transported via wormhole to the Irenas Penal Colony in the Plains region of Torden continent on the planet Arieki

August 18, 2008: Returned to the planet Foreas, beginning missions in the Plateau region of Valverde continent

August 19, 2008: Found the Vostok capsule used by General British (Richard Garriott) in the Divide region of Concordia

August 21, 2008: Encountered the Eloh named Gabriel in Eloh Sanctuary, teleporting from Concordia

August 22, 2008: Achieved experience level 30 and chose to follow the Exobiologist training course, rather than training in the Medic course

October 12, 2008: Launched with General British to the International Space Station; celebrated in the nightclub of the Staal Detention Center in the Ligo Crucible region on Arieki

October 15,2008: In Valverde on Foreas marshes spoke with the Eloh guardian, Jumna, who tells him to speak with his brother on Arieki; returns to Ligo continent on Arieki

October 16, 2008: Achieved experience level 40 at Boneyard Outpost in the Ashen Desert region of Ligo continent

October 31, 2008: Achieved experience level 50, the highest available, while killing beam manta creatures in the Thunderhead region of Ligo continent on Arieki

December 21, 2008: Gained the Earth Logos, the last available piece of wisdom provided by the ancient Benefactors, allowing him to return to Earth

February 27, 2009: Visited Earth for the last time, seeing the devastation in Madison Square Park, New York City

The references to *Specialist*, *Soldier* and other such terms concern decision points in the rather unique class and skill acquisition structure of *Tabula Rasa*. The quick-start interface card that came with the game also included a chart of the *character class tree*. Every avatar began in Recruit status, as yet with no specialization. Upon reaching level 5, the player faced a branch point, choosing to become a battle-oriented Soldier, or a technology-oriented Specialist. At each branch point, the player could save a clone of the avatar, returning later to experience the other option without needing to create a new avatar at the very beginning. I chose Specialist, and at level 15

Fig. 7.4 One avatar and five non-player characters in Staal Nightclub

needed to choose between Sapper, which really should have been called Engineer, and Biotechnician which I did in fact select. At level 30, a Biotechnician would become either a Medic supporting teammates in battle, or an Exobiologist, which I chose because it seemed more oriented toward research on extraterrestrial ecologies.

The Vostok capsule found in the Divide region of Concordia supposedly was left over from a second spaceflight by General British, in which he had been swept away from Earth to Foreas. When Richard Garriott carried all the *Tabula Rasa* player data into space with him, my avatar was on the other planet, Arieki, in the Crucible region of Ligo continent, an area torn by surging rivers of molten lava and the site of very disreputable institutions, including Staal Detention Center, Staal Junkyard, Thieves Quarter, Convict Pass, and Incurables Ward.

My avatar celebrated the real-world spaceflight of Richard Garriott in Staal Nightclub, shown in Fig. 7.4. He is the foreground figure dressed in battle armor, and the five other people are non-player characters who could always be found at the bar. They all appear to be brooding, and indeed most human NPCs in *Tabula Rasa* are devoid of spirit, other than in some cases the lust to kill. They did not want to leave Earth for Arieki, any more than the Brann criminals had wanted to leave Erdas, but were forced against their will by the Bane. One utopian group called the Cormans had reached Foreas before the attack on Earth, in an accidentally discovered alien spaceship, and are pacifists devoted to research. But aside from them, the humans generally seem stupid and vengeful—with the notable exception of the human players of *Tabula Rasa*.

For Halloween 2008, my avatar was able to wear a mask depicting the face of Richard Garriott, and had completed exploration of both Foreas and Arieki. Garriott's hope that *Tabula Rasa* would inspire many thousands of players with its visions of the human future in outer space was not realized. I must admit that I was much entranced by many of the ideas, including the catalog of reasons motivating interstellar travel by different groups. There was hope that a third alien planet would be added, and that more extensive crafting would add realism to the colonization effort. But that never happened. It may be that by the early twenty-first century, many of the people attracted inward to computer games were not inspired outward to the universe around us. Yet within *Tabula Rasa* lurked many philosophical ideas of much wider relevance.

7.4 The Logos Language

A chapter of Richard Garriott's autobiography describes his invention of the hieroglyphic Logos language of the Eloh for *Tabula Rasa*, calling it "a universal language." He had toyed with ancient Egyptian hieroglyphics, then studied Blissymbolics, invented by Charles Bliss, which Garriott called "a universal pictographic writing system that enables people to communicate with mentally handicapped children (Garriot and Fisher 2017)." A chemical engineer rather than a psychologist, Bliss was a survivor of the Buchenwald concentration camp in the Holocaust who came to believe that humanity needed a universal language to promote mutual understanding.[28] Inspired partly by Chinese writing, over a period of years he developed a system of symbols that had well-defined meanings but were not associated with any spoken language.[29] A group in Canada applied it to communication with disabled children (Wood et al. 1992). Although occasionally receiving some academic respect, the scientific status of claims about the value of Blissymbolics remains in doubt and its primary advocate today is a non-profit charitable organization named Blissymbolics Communication International.[30]

While inspired by Bliss, Garriott set out to devise his own set of symbols, for example using a simple drawing of an hourglass to represent *time*. Clearly, this is not a metaphor that could in fact be readily understood by aliens, let alone used as a medium of communication from one galactic civilization to another. Really, Garriott's goal was to develop a system of symbols that would become intelligible to players as they collected these Logos symbols from shrines across the two planets depicted in *Tabula Rasa*. Figure 7.5 shows my avatar standing before a pylon carrying an inscription in Logos symbols, intended to be read from top to bottom. The left side of the image shows the *tabula* itself, a part of the interface where the symbols

[28] en.wikipedia.org/wiki/Charles_K._Bliss, accessed May 2017.

[29] en.wikipedia.org/wiki/Blissymbols, accessed May 2017.

[30] www.blissymbolics.org/index.php/about-bci, accessed May 2017.

Fig. 7.5 Reading an Eloh inscription in a buried Forean temple

are placed after they have been collected, currently highlighting the time symbol, and other symbols that happen to be arranged in alphabetical order in terms of their English names.

The somewhat blurry Logos hieroglyphs on the pillar began with four that seemed to be in the tabula already, saying: "Time People Journey Cosmos." The dictionary in the commercial game manual provides a more complete set of meanings: "Time, We, Journey, Cosmos, To, Discover, Them, To, Give, Secret, Phi (Lummis 2007)." This might better be rendered, "It is time we journey into the cosmos to discover them and give the secret of Phi." In an earlier publication about *Tabula Rasa*, that also mentions the *StarCraft* interstellar strategy game, I explained the meaning of phi:

> Phi is a mathematical concept, also called the Golden Ratio, expressed as an irrational number, approximately 1.6180339887… but extending to an infinite number of decimal places. In some respects, therefore, it is comparable to the much better known irrational number, pi (3.14159265358979…). Phi can be generated mathematically in a number of ways, but among the most often cited is the ratio of two adjacent integers in a Fibonacci series, such as 1, 1, 2, 3, 5, 8, 13, 21… in which each later integer is the sum of its two predecessors. As the integers get larger, the ratio approaches ever closer to phi, but can only equal phi at infinity. Phi has been used extensively in both the arts and in mystical speculations for more than two thousand years. For example, in the *StarCraft* mythos, the discovery of phi opened the door to profound mental abilities for the Protoss, marking the dawn of their high-technology civilization (Golden 2007). As a metaphor, the Golden Ratio

suggests that the universe may contain hidden meaning that harmonizes with human hopes and perceptions (Bainbridge 2011b).

A considerable fraction of the pre-launch publicity for *Tabula Rasa* features the Logos language.[31] In a March 2007 interview, months before *Tabula Rasa* launched, Garriott explained that "advanced aliens that left a message throughout the universe would also likely use a pictographic language to make their messages easy to read by all civilizations that advanced far enough to develop language skills." He then explained its fundamental concept:

> The word "Logos" is used in our game both for the symbolic language and for the force-like super powers our players learn to wield. "Logos" is often used to describe god-like powers that manifest reality with the power of your mind and spoken words, such as "Let There Be Light and there was." Added to this is the concept of a logogrammatic language, which uses pictographic symbols for communication. Put them together and our game has both a language to learn whereby by the players will have a richer experience and be able to actually read writing in the art and architecture left behind by the Eloh in our game. It is also a great collecting game, where people in the game can collect these symbols to expand their personal dictionary. As the game progresses, players can use these Logos symbols to unlock special powers.[32]

Each logos in the tabula must be obtained from a different shrine, usually located in remote locations. The Logos obtained from the shrine depicted in Fig. 7.3 represents the concept *mother*, and one representing *father* could be obtained some distance south-southwest of it in Uyona Chasm. In the case of a few very important symbols, a cavern containing a shrine could not be entered unless one had already collected the other symbols required to give it philosophical meaning. For example, the powerful *vortex* symbol could be found in the Plateau region of Valverde continent on the planet Foreas only after collecting symbols that could spell out this mystical sentence: "Eloh Empower Only The Strong." Notice that the words in this sentence belong to different grammatical categories. Garriott acknowledges that some linguistic abstractions were difficult to represent graphically, and required him to use cartoon figures of tiny people:

> My greatest challenge was how to communicate personal pronouns. The basic symbol I used consisted of two connected figures, meaning us, we, them, or our rather than I. I decided that a single quote mark above the figure means first person, I or me, two marks means second person, we or us, and third person, three marks, is them. A house obviously is a house or shelter, while two houses represent a community or a civilization. A very small person lying down is a baby, and a standing small person is a child (Garriot and Fisher 2017).

Figure 7.6 shows six sentences transcribed in Eloh Logos, selected because they provide the best introduction. The first five sentences read downward, just as on the pylon in Fig. 7.5, while the sixth goes horizontally from left to right. All the symbols needed to be collected, in order for my exobiologist avatar to gain all his

---

[31] Alexander, Justin. 2017. Tabula Rasa: The Language of Logos. April 14th, 2007, http://thealexandrian.net/wordpress/614/random/tabula-rasa-the-language-of-logos;http://www.engadget.com/2008/10/14/richard-garriotts-logos-message-to-tabula-rasa-players-decipher

[32] www.gamesindustry.biz/articles/tabula-rasa, accessed May 2017.

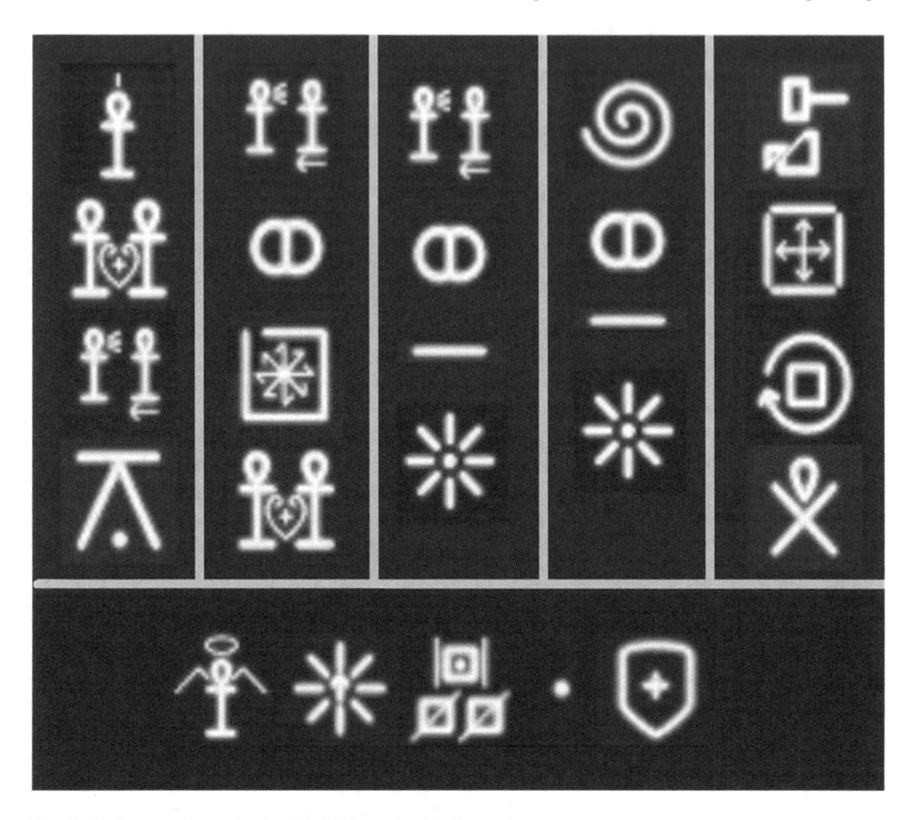

Fig. 7.6 Six sentences in the Eloh hieroglyphic language

fundamental technological abilities. The symbols in the first sentence represent: Self, Friend, Summon, Here. The first one, *self*, is the figure of a person with one mark over its head, the expression for first person pronoun explained in the above Garriott quotation. The second symbol shows two people standing with a heart between them that contains a plus sign, meaning *friend*. The third symbol, for *summon*, depicts one person speaking to another, and that other moving toward the first as suggested by the arrow under the feet. The final symbol in the sentence is based on Garriott's symbol for location, which means *here* if a dot is placed at the bottom, and *there* if the dot is placed at the top. I suspect it represents a road leading to the horizon. Once my exobiologist had all four of these symbols, he earned the interface icon for Create Clone, which creates a battle-ready twin of the avatar to serve as his companion.

The second Eloh Logos sentence begins with the glyph for *summon*, and ends with *friend*, so it expresses a similar meaning to that of the first sentence, gaining a "friend" or at least something beneficial. It is the sentence that gives an exobiologist the "hortimonculus" power, the ability to make a plant radiating a healing aura to sprout from a corpse, which could be life-saving in a pitched battle. The second symbol represents *life* as two cells of an organism, perhaps at the point when one cell divides. The third shows a square box containing something like an explosion and means *control*. So the sentence that confers the hortimonculus power is: Summon, Life, Control, Friend.

The third sentence begins with *summon friend*, then adds a minus sign which means *negative* or *subtract* and a starlike radiation of eight lines with a dot at the center, meaning *spirit*. This confers the reanimation ability, reviving one enemy corpse briefly as an ally: Summon, Life, Negative, Spirit. Interestingly, a symbol found only in the commercial guidebook is identical to *spirit*, but with a minus sign in the center rather than a dot, and means *secular*, while a plus sign in the center gives an almost identical glyph the meaning of *supernatural*. The fourth sentence is identical to the third, except starting with a spiral meaning *vortex* and conveying a more powerful version of the same ability, a reanimation wave that causes all nearby enemy corpses to revive temporarily: Vortex, Life, Negative, Spirit.

The fifth sentence gives the exobiologist the cadaver immolation ability, that causes a corpse to explode after a brief delay, giving the avatar a chance to stand back and entice an enemy to advance near to the corpse: Damage, Area, Around, Death. The first of these four shows a hammer doing *damage* to a square by knocking a corner off it. The *area* glyph uses arrows inside a square to indicate its surface. The *around* glyph could hardly be more obvious, an arrow going around a square. To me the glyph representing death looks like a person who has been crossed out, but I do not find analogies in the tabula to support this interpretation.

The sentence of five glyphs arranged horizontally at the bottom of Fig. 7.6 is the set that are required to unlock the shrine where the *vortex* glyph can be found: Eloh, Empower, Only, The, Strong. The first, meaning *Eloh*, shows a person with halo and wings, clearly an angel. The glyph for *empower* is like spirit but with an upward-pointing arrow in the center. The *only* glyph crosses out two of three squares and selects the one at its top as the only one of interest. A simple dot indicates *the*. The *strong* glyph is a shield with a plus sign in it, and another glyph in the dictionary represents *weak* by placing a minus sign instead.

To return to the point that Garriott's hieroglyphic language is culture-bound yet gives users the sense that they are interpreting an alien language, we can note some of the terrestrial conventions built into these symbols. Affection, as in *friend* but also in *love*, is represented by a conventional symbol for the human heart, actually not very realistic and probably quite different in alien species. As is true through many of the virtual worlds explored in this book, sentient beings are assumed to have two arms, two legs, and one head. Had Garriott been a native speaker of a language that ordinarily uses abstractions from ancient pictograms, notably Chinese or Japanese, he might have employed different principles to devise his glyphs.

A classic if extremely obscure science fiction example of cultural bias is found in a map of the planet Venus included by Edgar Rice Burroughs in some editions of his novels about that planet. He devised an exotic alphabet for the inhabitants of that planet in which a long "u" (as in the English word tulip) is represented as ōō, perhaps forgetting that English is practically unique in assigning this pronunciation to double-o.[33] An example from Fig. 7.6 is the word *the*, the English definite article, if we recognize that many terrestrial languages lack *the*, as Wikipedia explains: "Articles are found in many Indo-European languages, especially Romance languages, Semitic languages (only the definite article), and Polynesian languages, but are formally absent from many of the world's major languages, such as Chinese, Indonesian, Japanese, Hindi, Punjabi, Urdu, Russian, the majority of Slavic and Baltic languages, Yoruba, and the Bantu languages. In some languages that do have articles, like for example some North Caucasian languages, the use of articles is optional but in others like English and German it is mandatory in all cases."[34]

It is worth noting that the official game manual and the commercial guidebook spell the sentences for *reanimation* and *reanimation wave* differently from each other, and from the versions in Fig. 7.6 which were taken from in-game screenshots. The official manual used a cartoon of a stick figure upside-down, representing the concept *corpse*, instead of the glyphs for *negative spirit* and added the *friend* glyph to *reanimation* to form: Summon, Life, Corpse, Friend. In contrast, reanimation wave required only three glyphs: Vortex, Life, Corpse (Garriott 2007). In the commercial guidebook, the sign for negative is a vertical line with a narrow underline, defined as *no or negatory*, and the glyph in Fig. 7.6 is defined simply as *subtract*. Both books were printed prior to the game's release, and the guidebook implies that the full set of glyphs was intended to be found in the game, perhaps on additional planets that never were created. Neither the corpse nor negatory glyphs could in fact be found in shrines on either Foreas or Arieki.

7.5 Conclusion

In Garriott's cosmic history, Humans were refugees from an Earth that had been stolen from them, while the Foreans fled their home planet because they, themselves, had ruined it. The Brann on Arieki were descendants of prisoners exiled by members of their own species, while the Thrax were slaves of the Neph. Perhaps only the Eloh experienced interstellar travel in joy, motivated by the glory of discovery, and the virtue of sharing knowledge. Apparently few real-life Humans are *receptives*, capable of adopting the Eloh dream, because *Tabula Rasa* gained too few subscribers to survive in the gamer marketplace. Perhaps its intellectual merit was academic, appropriate for an intelligent, sophisticated audience, at a time in cultural history when the kinds of economic and social institutions did not yet exist, to support such an endeavor.

What evolutionary steps would be required for humanity to evolve into Eloh? Perhaps we need wait only a few millennia or geological eras, until our brains and societies had naturally risen to the Eloh stage. Or, a radical social movement, comparable to a millenarian religion, could inspire us with the necessary vision. One

[33]en.wikipedia.org/wiki/Amtor, www.erbzine.com/mag30/3040.html, accessed May 2017.

[34]en.wikipedia.org/wiki/Article_(grammar).

candidate for that transformative role does currently exist, namely Transhumanism, that advocates transformation of human nature through the application of biological enigineering and advanced artificial intelligence (FM-2 030 1989; More and Vita-More 2013; Bainbridge; 2017). Yet the Tranhumanist Movement is controversial on several gounds. Some believe that it is evil precisely because it works against human nature (Fukuyama 2002, 2004). Its leading members seem to function intellectually as philosophers adressing the ethics issues connected to human transcedence, rather than scientists and engineers prepared to achieve it. However problematic the idea may be, it is possible that humanity will not be able or motivated to expand outward from Earth until it had approached Garriott's vision for the Eloh.

Tabula Rasa could contribute to realization of that vision, if it could be revived, but that idea raises the more general question of how to preserve virtual worlds—artworks distinctive to our time—from destruction. We could imagine NCSoft placing *Tabula Rasa* in a kind-of public virtual world library, where it would forever be available to college classes, researchers, and a few individuals willing to pay a modest subscription price to access all the worlds in the dynamic archive (Rumilisoun (William Sims Bainbridge) 2010). If and when real space emigration begins, we may also imagine the archive could be transported to Mars and other colonies, as a basis for the distinctive cultures that will evolve on those other worlds, including within it all the spaceflight-related simulations explored in this book.

References

Backe, Hans-Joachim, and Espen Aarseth. 2013. Ludic Zombies: An Examination of Zombieism in Games. In *Proceedings of DiGRA 2013: DeFragging Game Studies*. Finland: The Digital Games Research Association.

Bainbridge, William Sims. 2011. *The Virtual Future,* 41, 46. London: Springer.

Bainbridge, William Sims. 2013. *eGods: Faith Versus Fantasy in Computer Gaming*, 279. New York: Oxford University Press.

Bainbridge, William Sims. 2014. *Personality Capture and Emulation*, 38–42. London: Springer.

Bainbridge, William Sims. 2017. *Dynamic Secularization*. London: Springer.

FM-2030. 1989. *Are You a Transhuman?* New York: Warner.

Fukuyama, Francis. 2002. *Our Posthuman Future: Consequences of the Biotechnology Revolution*. New York: Farrar, Straus, and Giroux.

Fukuyama, Francis. (2004, September–October). *Transhumanism. Foreign Policy*.

Garriot, Richard, and David Fisher. 2017a. *Explore/Create: My Life in Pursuit of New Frontiers, Hidden Worlds, and the Creative Spark*. New York: William Morrow.

Garriot, Richard, and David Fisher. 2017b. *Explore/Create: My Life in Pursuit of New Frontiers, Hidden Worlds, and the Creative Spark*, 85. New York: William Morrow.

Garriot, Richard, and David Fisher. 2017c. *Explore/Create: My Life in Pursuit of New Frontiers, Hidden Worlds, and the Creative Spark*, 87. New York: William Morrow.

Garriott, Richard (ed). 2007. *Welcome to the AFS: Allied Free Sentients Official Field Manual*, 55. Austin, Texas: Destination Games.

Golden, Christie. 2007. *Firstborn*, 92, 232. New York: Pocket Books.

Lummis, Michael. 2007. *Richard Garriot's Tabula Rasa*, 222, 234–248. Indianapolis, Indiana: BradyGAMES/D.

More, Max, and Natasha Vita-More. 2013. *The Transhumanist Reader: Classical and Contemporary Essays on the Science, Technology, and Philosophy of the Human Future*, 2013. New York: Wiley-Blackwell.

Royce, Bree. 2017. Massively OP's Ultima Online 2016 tour. http://www.youtube.com/watch?v=zepWGUhA3q8 Accessed May 2017.

Rumilisoun (William Sims Bainbridge). 2010. Rebirth of Worlds. *Communications of the ACM 53*(12): 128.

Wolfe, Tom. 1979. *The Right Stuff*, 17. New York: Picador.

Wood, Claudia, Jinny Storr, and Peter A. Reich. 1992. *Blissymbol Reference Guide*. Toronto: Blissymbolics Communication International.

Chapter 8
A Virtual Human-Centered Galaxy

Over the past four decades, the phenomenally popular *Star Wars* franchise has significantly shaped the popular impression of the human future in outer space, and some of the videogames count as computer simulations of social and cultural interaction beyond the bounds of Earth. For all their cosmic popularity, these movies and games usually do a very poor job of presenting the realities of space travel and other worlds. Yet simulations need not always be physically realistic in order to illustrate sociological theories or explore possible correlations between humanly-meaningful variables. While this chapter will be critical of some technical and cultural aspects of *Star Wars* computer games, it will seek interesting lessons in components of them that may be valuable as social simulations of space exploration, or express the hopes that ordinary people may have about the human future in space. Thus it is extremely significant that the *Star Wars* universe is very different from the one we apparently inhabit, filled with emotion, social conflict, and meaningful human action rather than vast emptiness.

8.1 A Vast, Imaginative Tradition

At the beginning of 2017, Wikipedia listed fully 95 different *Star Wars* videogames and computer games that had been published.[1] Two were inherently social and exceedingly complex massively multiplayer online role-playing games, *Star Wars Galaxies* which prevailed from 2003 until 2011, and *Star Wars: The Old Republic* (SWTOR), which launched immediately after the demise of *Galaxies* and will be the main focus of this chapter. Another Wikipedia page reports: "In 1978, Apple Computer produced an unlicensed Star Wars game on cassette tape for its Apple II. As a 'space pilot trainee', the player destroys TIE fighters using a first-person heads-up display. The first video game cartridge bearing the name Star Wars appeared that

[1] en.wikipedia.org/wiki/List_of_Star_Wars_video_games, accessed February 2017.

W. S. Bainbridge, *Computer Simulations of Space Societies*, Space and Society,
https://doi.org/10.1007/978-3-319-90560-0_8

year on the RCA Studio II clones Sheen M1200 and Mustang Telespiel Computer."[2] The real beginning for systems like those used today was 1982 and a game for the Atari 2600 based on *The Empire Strikes Back*, while my own introduction to the genre was the 1992 Super Nintendo videogame *Super Star Wars*.

It would be feasible to obtain the old hardware and explore these games in full detail, but their social aspects are exceedingly limited in most cases, so an efficient alternative is to examine YouTube videos that players have posted of their own experience. By placing a record of their own actions in a public video archive, players rendered these solo games somewhat social, just as Facebook groups did for *No Man's Sky*. In addition, scholars may consult reviews and other publications concerning specific games, assisted by the Wikipedia pages devoted to many of them.

Several brief YouTube videos illustrated the 1982 game for the Atari 2600, which has very, very simple two-dimensional graphics. The player is operating a small fighter aircraft called a *snowspeeder* to attack huge walking battle tanks called AT-ATs (All Terrain Armored Transports), as in one scene of the second movie, *The Empire Strikes Back*.[3] The snowspeeder is represented by tiny rectangles, and the player seems to have control only over its movement up and down or left and right, and firing a gun at the AT-ATs. This is a primitive version of what is called a *side-scrolling* game. The snowspeeder is always near the center of the image, moving merely up and down, while the scene moves left or right to represent the player flying. AT-ATs appear at the left, shooting at the player, who goes up or down to avoid their shots while firing at them. The player can escape by flying to the right.

This does not seem social at all, yet combat is a form of interaction, and the rudiments of a story add complexity. Five AT-ATs are attempting to walk far to the right, and the player need not simply destroy them one at a time. An alternative is to damage the first AT-AT to slow it and the others down, then fly over it and destroy the second AT-AT, and so forth. The AT-ATs have a goal of walking far to the right to destroy Echo Base, which belongs to the Republic, which is at war with the Empire, which owns the AT-ATs. Occasionally a simple version of the *Star Wars* tune plays to inform the player that temporarily the mystical Force will protect the snowspeeder from the missiles fired at it by the AT-ATs. A couple more technical details add further temporary options, thus requiring the player to learn and to think, rather than just to shoot wildly while dodging the enemy's missiles. An especially thoughtful 2006 YouTube video, posted by an expert gamer calling himself "tr0d," outlines these subtleties and had been viewed 234,400 times over the following decade.[4]

The complexity of electronic games, in the sophistication of their stories as well as graphics, has depended upon the computational capabilities of their hardware. Sold over the remarkably wide span of years 1990 until 2003, the Super Nintendo

---

[2]en.wikipedia.org/wiki/List_of_Star_Wars_video_games, accessed February 2017.

[3]en.wikipedia.org/wiki/Star_Wars:_The_Empire_Strikes_Back_(1982_video_game); en.wikipedia.org/wiki/Walker_(Star_Wars)#All_Terrain_Armored_Transport_.28AT-AT.29, accessed February 2017.

[4]tr0d, "Star Wars (Atari 2600) (How To Beat Home Video Games 1)," June 15, 2006, www.youtube.com/watch?v=6SkWIPKFrxk, accessed February 2017.

Entertainment System was for a time the dominant system, and *Super Star Wars* had considerably improved graphics despite also being a two-dimensional side-scroller in most of its short episodes. Like many of the games of its period, it was divided into a series of *stages*, many of which typically ended with a *boss battle*. For example, the first stage depicts Luke Skywalker running to the right across the desert planet Tatooine, shooting a pistol at various monsters, ending with a long battle against the Sarlaac, which has its own Wikipedia page that describes it thus: "a multi-tentacled alien beast whose immense, gaping maw is lined with several rows of sharp teeth."[5]

Much of *Super Star Wars* is a *platform game*, in which the player's character must jump from one surface to another, often unstable or moving, while killing enemies but without interacting in complex ways with any but the stage bosses. As was common for popular-culture games of the period, it was possible to select among characters to play, whether Luke Skywalker, Han Solo, or Chewbacca. A very indirect social aspect concerned the environment in which a particular stage was set, which sometimes was a distinctive technological locale like the interior of the Death Star battle station. One of the many bosses is the tractor beam device which must be deactivated in order for the Millennium Falcon spaceship to escape the Death Star. The climactic stage involves flying a starfighter spacecraft down the trench on the surface of the Death Star to deliver the missile that will destroy it, thus representing one kind of spaceflight and one rather rare mode of hostile social interaction. Currently, videos of the entire game are available for online viewing and detailed analysis.[6]

When *Star Wars Galaxies* launched in 2003, ordinary desktop computers had good enough graphics that a television set was not needed as the display device, which was the case for the Super Nintendo. The author explored this virtual world for 618 h, and published three book chapters about it, analyzing a number of ways in which it was a realistic simulation of the universe depicted in the movies, if not of the universe we actually inhabit (Bainbridge 2011; 2016). Several planets could be visited, all offering interesting three-dimensional environments in which resources could be gathered and adventures experienced. All of the planets were representations of terrestrial environments, in some cases exaggerated, such as the forest planet Kashyyyk, which is the origin of Wookiees like Han Solo's friend Chewbacca, where much of the action takes place far above ground on immense tree branches.

On seven of the planets, players could build homes, cluster them in villages, and set up shared resources like factories and public transportation hubs called *shuttleports*. Figure 8.1 shows my main Jedi knight avatar, named Simula Tion, in the living room of the home she shared with an engineer named Algorithm. She has put pictures on the walls and a carpet on the floor. The two main objects to the right of her were standard sculptures of the symbols of the Republic on the left, and Empire on the right, that were designed to be outdoors in a public space, but fitted fine in her living room. Behind the right wall was the workshop where Algorithm could craft droids

---

[5]en.wikipedia.org/wiki/Sarlacc, accessed February 2017.

[6]www.youtube.com/watch?v=NnsGbd8WMNg; www.youtube.com/watch?v=bb39BV6Gs40&t=835s, accessed February 2017.

Fig. 8.1 The living room of a virtual house on the planet Tatooine

and other equipment using big machines. Another room on this floor was used for storage, including chests where many virtual objects could be placed, and there also were smaller rooms upstairs.

The lock on the front door of a home like this could be set to permit friends to enter. Many inhabitants of *Star Wars Galaxies* allowed anyone to visit, using their homes as museums to display all the awards and artifacts they had collected and furniture, clothing or sculptures they had created. To make an object required gaining a specific skill and collecting raw materials, which could be purchased from other players, shared among friends, but ultimately gathered from the natural environment or from the corpses of monsters and enemies. Clearly, players made very significant investments of time, which they undoubtedly enjoyed, but that were lost forever when *Star Wars Galaxies* was shut down at the end of 2011.

In mid-September of that year, I used the in-game data interface to estimate the extent of the virtual real estate that players had developed. At that point, separate simulations existed on four Internet servers, each called a *galaxy*: Chilastra, FarStar, Flurry, and Starsider. So I sent an avatar in each of those galaxies to each of the seven habitable planets, to gather the data summarized in Table 8.1. Each shuttleport represented a village of player homes, and the factional bases were rallying points for teams of players in the conflict between the Republic and the Empire.

Table 8.1 Player-created group infrastructure on seven simulated planets

Planet	Players' shuttleports				Players' factional bases			
	FarStar	Starsider	Chilastra	Flurry	FarStar	Starsider	Chilastra	Flurry
Tatooine	10	10	9	9	25	20	14	23
Naboo	9	10	10	7	26	25	16	28
Corellia	6	10	6	6	19	16	11	25
Dantooine	18	18	19	17	22	24	12	17
Lok	20	20	12	20	20	22	14	17
Rori	9	18	8	10	27	24	12	22
Talus	14	20	20	10	22	21	13	25
Total	86	106	84	79	161	152	92	157
Percent Republic	–	–	–	–	35%	53%	70%	52%

The main thing to notice about Table 8.1 is that in fact groups of players had created an objectively large number of ports and bases. Many shuttleports were across a town square from a city hall and other facilities players had built, although many towns did not bother to construct a shuttleport because they were situated near public transportation. When *Star Wars Galaxies* ended its life, many players migrated to *Star Wars: The Old Republic*, but they could not move their virtual property with them. Thus, like any laboratory experiment, this simulation of seven inhabited planets came to an end.

8.2 Far from the Bright Center

A good starting point for understanding the new universe simulated by *Star Wars: The Old Republic* is the planet where the first movie began, Tatooine. In an earlier book, I had devoted a chapter to the experience of living there, as presented in *Star Wars Galaxies*:

> When C-3PO asked what planet he was on, Luke Skywalker replied, "Well, if there's a bright center to the universe, you're on the planet that it's farthest from."[7] Named Tatooine, it was a desert world where the farmers' main crop was water, painfully harvested from the dry atmosphere. A frequent location for the action in *Star Wars* movies, it contained the boyhood homes of both Luke Skywalker and his father Darth Vader, as well as the haunts of Jabba the criminal boss and Watto the junk dealer. Tatooine was the first planet visited by players of *Star Wars Galaxies* (SWG), and many felt it was their virtual home. *Star Wars Galaxies* was noteworthy among popular MMOs in the extent to which it encouraged people to experience social life freely in a virtual environment, rather than follow a rigid course of story-based missions that were predefined by the game designers (Ducheneaut and Moore 2004; Ducheneaut et al. 2007).

[7]Lucas, George. 1976. *StarWars: A NewHope,* script. http://www.blueharvest.net/scoops/anh-script.shtml, accessed 15 Jan 1976

Star Wars: The Old Republic also portrayed Tatooine, but players could not build homes, factories, and towns there, as they could in *Star Wars Galaxies*. In the same year that *Galaxies* was killed, and *The Old Republic* was born, an atlas of SWTOR's initial planets was published, called *Explorer's Guide*. It introduced Tatooine in these words:

> Far in the Outer Rim, the sands of Tatooine bake beneath the glare of two bright suns. Small pockets of barely civilized communities dot the desolate landscape, surrounded by the endless expanse of barren dunes and rocky canyons that have silently slain so many of those who ventured out into the desert. Among the small shantytowns and settlements that persist, travelers may find shelter from the brutal climate, but trust is as rare as water on this lawless world. Visitors and locals alike must constantly watch their backs in Tatooine's townships (Searle 2011).

We may well ask what these two simulated versions of a distant planet represent? Historically, they are derived from the perspective on the planet Mars held by astronomers a century before SWTOR was launched in 2011. It has been well documented that George Lucas invented the *Star Wars* saga only after he failed to get the rights to produce a *Flash Gordon* movie, and the *Flash Gordon* saga was created only when a publisher of comics failed to get the rights to adapt the Mars novels of Edgar Rice Burroughs, who wrote his first book in 1911, greatly inspired by the popular writings about Mars by astronomer Percival Lowell (Gold 1988; Barrett 2004; Rinzer 2007; Lowell 1906; Lowell 1908). More abstractly, Tatooine represents planets where human colonization is just barely possible, without the need for really radical technological innovations.

The lawless nature of Tatooine society might reflect the cowboy mythology of the Wild West of nineteenth-century America, but for our purposes here it reflects the fundamental theoretical issues about how social order may arise from chaos. In *Star Wars Galaxies*, groups of players were encouraged to create their own cohesive communities, but that is far more difficult in SWTOR, not the least because it emulates novels and movies very strictly, in requiring the user to play a largely pre-scripted role in a pre-written story. We will illustrate how SWTOR does that, by very briefly summarizing the stories in which two different characters visit Tatooine in the middle of their story arcs, Hysterion the Jedi consular and Sociopathy the Sith warrior. Their names were chosen by their player, but their stories were largely predetermined by the game designers.

Once upon a time, indeed many years before the historical period covered by the *Star Wars* movies, an entire galaxy suffered under social disorganization, after war between fanatics favoring a Republic versus an Empire had ground to an indecisive halt. At the historical moment the first *Star Wars* movie depicted, both galactic factions had abandoned Tatooine, but SWTOR depicts an earlier time when each held a competing spaceport and a scattering of smaller bases. Environmentally and technologically, the planet was almost identical, however, for example arid and requiring "moisture vaporators" to extract water from the air, there being no rivers or ponds. In the years after the early Mars novels of Edgar Rice Burroughs, astronomers had discovered there were in fact no canals on Mars, but his stories included the model for moisture vaporators, because atmosphere factories needed to run full tilt every

Fig. 8.2 A transportation hub at the imperial headquarters on the planet Tatooine

day to replenish oxygen to the atmosphere of a rusting planet. Figure 8.2 shows a scene on Tatooine that could just as well have been Mars, except there is oxygen in the air to breathe, and the sands are not so rusty.

Figure 8.2 is a scene at the outskirts of the Imperial spaceport on Tatooine, named Mos Ila, comparable to Mos Eisley which features in the first *Star Wars* movie. The corresponding Republic spaceport, far to the north, is Anchorhead, mentioned but not shown in that same film. In the foreground, Sociopathy the Sith warrior is riding an Eternal Empire Patroller, a *speeder* comparable to a motorcycle but lacking wheels and levitating above the sand through some mysterious technology. Immediately behind him, a human woman is negotiating with two Jawas assisted by a droid, probably to lease the tame dewback animal behind her, for some mission she must perform. Behind them is the facility where avatars may pay to ride a speeder along a pre-set route to one of the other speeder locations on Tatooine.

Thus this picture shows three kinds of rapid transit, better than walking over long distances: (1) public transportation to pre-defined destinations, (2) riding one's own vehicle, and (3) riding an extraterrestrial beast. Later, we will consider the personal vehicles more closely, but here are descriptions of dewbacks from two online SWTOR encyclopedias:

> Dewbacks were Tatooine reptiles that could be used as mounts. Specialised Sandtroopers called Dewback troopers used them while searching for the stolen Death Star plans in the planet's deserts. Female dewbacks laid 50–85 eggs each year. Dewbacks got their name because they lick morning dew off their backs.[8]

> Dewbacks were large, four-legged, omnivorous, cold-blooded reptiles native to the Dune Sea of Tatooine. As such, they were well-adapted to the harsh desert climate. This and the fact

[8]starwars.wikia.com/wiki/Dewback, accessed February 2017.

> that they could be easily domesticated, made them commonly employed as beasts of burden by both the inhabitants of the desert planet and the off-worlders wandering its dunes.[9]

Sociopathy was the apprentice of a very evil senior warrior, Darth Baras, who would stop at nothing to destroy his own teacher and gain a dominant position on the Dark Council that rules the Empire. Early missions required eliminating some of the spies who have been serving Darth Baras, which means betrayal of loyal betrayers, a nearly infinite regress of deceptions. The mission that sends a warrior to Tatooine begins a quest to locate Jaesa Willsaam, an apprentice to a Jedi Knight. In principle, according to *Star Wars* mythology, the republican Jedi follow the light side of the Force, and the imperial Sith, the dark side. Yet SWTOR allows each avatar to move toward either light or dark, regardless of their faction, by performing actions more in tune with one or the other. Long after leaving Tatooine, Sociopathy was not only able to locate Jaesa Willsaam, but also to recruit her to become one of his five companions. In becoming the follower of a Sith warrior, Jaesa performs an apparent act of treason against the Jedi, yet as an SWTOR wiki explains, she may do this from either a light or dark perspective:

> Light Side:
>
> Having now seen the darkness within the Jedi Order, Jaesa has left its corruption behind completely to help transform the Empire from within. She has searched her whole life for something to believe in, and her new teacher has provided the answer.
>
> Dark Side:
>
> Having finally witnessed the Jedi Order's weakness and the dark side's true power, Jaesa embraces the Sith path with reckless abandon. She now knows that the only truth-inducing force in the galaxy is fear.[10]

Thus, Jaesa is a simulated person, with private psychological issues, and even some supernatural qualities because she is capable of using the power of the Force. From the first *Star Wars* movie onward, very few people and absolutely no droids had the spiritual sensitivity required for the Force, yet as a computer-generated character, Jaesa could be classified as a droid in human form. We could classify her as a secondary avatar, rather than non-player character, because she serves the goals of the player, but does so standing near the primary avatar. Given the Jedi training she has received, she can use a light saber as her weapon, and she can use Force magic to heal the primary Sith avatar when he is wounded. In SWTOR, a complex social relationship develops between each companion and the avatar, illustrated in conversations they have, as well as often when the companion expresses a judgment about the avatar's recent behavior. Thus, SWTOR simulates social relations among human beings within a civilization rather more than it does astrophysics within a galaxy.

Hysterion the Jedi consular visited Tatooine as one of several planets where a senior Jedi was suffering from an ancient, spiritual disease called the Dark Plague, that turned them covertly from the light to the dark side. To remain on the light

[9]torf.mmo-fashion.com/jundland-dewback, accessed February 2017.

[10]swtor.wikia.com/wiki/Jaesa_Willsaam, accessed February 2017.

side of the Force, he was required to perform powerful healing magic on each, which weakened his own powers dangerously. Thus, playing the consular story in dedication to the light side effectively gives the player the personal experience of becoming a virtual Christ, providing salvation to others through self-sacrifice.

Although SWTOR players cannot construct factories and houses on Tatooine, as they could in *Star Wars Galaxies*, they may purchase a *stronghold*, which is an estate located at some remote desert location. After obtaining the first section, other installments may be purchased over time. Like Luke Skywalker's boyhood home, the entrances are set below ground level on a sandy courtyard, to protect the inhabitants from the sting of wind-blown sand. On the large upper floor, called the *cliffside balcony*, decorations may be placed at many locations in either of two rooms. The player goes into edit mode, which displays many squares and rectangles on the floor, where a virtual object may be placed, such as a statue of the feeble protocol droid 2 V-R8 that is usually found on the spaceship of any Republic avatar, or of a powerful female robot named Scorpio who considers herself far superior to any human being. The main building consists of two stories, connected by an elevator, and two additional rooms may be added later. Four homestead rooms with their own entrances are also available. The extensive exterior includes a moisture vaporator and a landing pad for spaceships where a docking bay may also be bought.

There are two kinds of money in SWTOR, credits and cartel coins. Credits, a term common in science fiction, are the in-game currency earned by an avatar for completing missions, looted off dead enemies, and earned by selling valuable objects and raw material to non-player vendors. Cartel coins are bought for dollars, the cheapest rate being 5,500 coins for $39.99 or about 138 coins per dollar, early in 2017, and 500 coins per month are given to players who have fully subscribed. The purchase price of the Tatooine stronghold was 2,500,000 credits or 2,500 cartel coins. Each of the four homestead rooms costs 900,000 credits or 950 cartel coins, the two cliffside rooms cost 800,000 credits or 900 coins each, while the docking bay cost 1,800,000 credits or 2,000 coins. Items to decorate the rooms and some outdoor yards of a stronghold can be purchased from the game company by opening a store interface inside SWTOR and paying cartel coins, or from other players with credits in an auction system called the Galactic Trade Network. A stronghold can be visited by all of the avatars on the particular account, which are on the particular one of SWTOR's 17 Internet servers, and by the avatars of other players who are invited to do so.

Through a census to be described more fully in the following section, fully 1,305 people were on the planet Tatooine in the middle of Sunday evening, January 29, 2017, local time. That is to say, an avatar of mine in each of the two factions on each of the 17 SWTOR servers checked the "who" part of the team-formation part of the game's interface, specifying "Tatooine." This gave the population of people, not merely avatars, because the census was taken quickly so that users had no time to switch to second or third avatars, at least not on a single server, and cross-server traffic tends to be low. Of these 1,305 people, 754 or 57.8%, were currently running an avatar that belonged to the Empire faction, while only 511 or 42.2% had allegiance with the Republic. This is a general finding from SWTOR censuses: Although the heroes of the movies belong to the Republic, in multiplayer games the Empire is more popular.

The census tells us more about the players, because the servers are not identical to each other. Nine of the 17 servers are for European players, three each in English, French, and German. Of the total 613 SWTOR people in Europe, 108 or 17.6% were on French-speaking servers, and 172 or 28.1% were on German-speaking ones. The 333 or 54.3% on European English-speaking servers presumably included Europeans with many native languages who have some knowledge of English, and indeed online games are among the environments where people can improve their knowledge of this "universal language," ironic as it may be to call English the "lingua franca." It is possible for players to run avatars on all of the servers, as for example I have done, so some international players may often use the 8 North American servers, where Tatooine had a population of 692 or 53.0% of the total 1,305.

A nearly universal design decision for massively multiplayer online games is how to structure direct combat between avatars. Much of the combat take place between avatars operated by people and non-player characters (NPCs), such as monsters or the warriors of fictional factions, operated purely by algorithms in the form of simple artificial intelligence. One design option is to set aside special subareas of a virtual world, often called *arenas* or *battlegrounds*, where the programming allows players to fight each other, often in teams. In recent years, some games like *League of Legends* have become very popular without offering anything other than a battleground.[11] SWTOR offers battlegrounds, but it also has the more traditional design solution of servers set aside for *player-versus-player* (PvP) combat. Within a PvP world, some of the areas are safe, notably the home planets for the two factions, but outside them the avatar of one player may attack the avatar of another player that belongs to the opposing faction. Given the common term PvP, the normal servers are often called PvE for *player-versus-environment*. Only 102 or 7.8% of the 1,305 people on Tatooine at the time of the census were on PvP servers.

While the difference between PvP and PvE is determined by the programming, another behaviorally significant distinction is determined by culture. Some servers are described as *role-playing* or RP, which means that players are encouraged to stay within the dramatic role of their avatars. For example, they may type into the text chat shared with other players things that their avatar might logically say, and groups of players may set up special role-playing events that expand upon the story content programmed into the game. If they want to communicate something about their offline lives, they should ideally preface it with "OOC" which stands for *out-of-character*. All but one of the 6 RP servers is also PvE, and the one RP PvP server had only 9 people on Tatooine when the census was done. The total following the RP norms was 501 or 38.4% of the Tatooine population.

[11] en.wikipedia.org/wiki/League_of_Legends, accessed February 2017.

Table 8.2 Census of origin planets and politico-economic hubs

	Function	Region	Parsecs	Population
REPUBLIC				
Ord Mantell	Non-force origin	Coreward worlds	23,969	494
Tython	Jedi origin	Coreward worlds	19,143	1,167
Coruscant	Government	Coreward worlds	21,104	1,404
Fleet	Social-economic hub	Coreward worlds	22,324	1,401
EMPIRE				
Hutta	Non-force origin	Hutt space	12,672	528
Korriban	Sith origin	Seat of the empire	29,012	1,096
Dromund Kaas	Government	Seat of the empire	30,318	2,246
Fleet	Social-economic hub	Seat of the empire	28,805	1,856

8.3 A Simulated Galaxy

Tatooine is but one of many worlds in the SWTOR galaxy, 27 of which can be visited repeatedly and extensively explored.[12] Indeed, they are conceptually distributed across a full spiral galaxy depicted in the computer interface of each spaceship as a player selects the next destination. Near the center is a region called Coreward Worlds, where the Republic holds sway. Tatooine is in the remote region called Distant Outer Rim. Almost on the other side of the galaxy from Tatooine is the Seat of the Empire. The earlier war between the two factions left the Republic in control of the heart of galactic civilization, but struggling to survive amid the ruins of its destruction, while the Empire had retreated and was obsessed with regaining a central position. Most planets were relatively lawless, but one other multi-world faction existed, under somewhat precarious control by the Hutts, gigantic intelligent slugs who are criminal overlords, like their representative Jabba the Hutt on Tatooine in the original movies.[13] Table 8.2 lists the six worlds and two "fleet" space stations exclusively used by one or the other of the factions, along with their distances in parsecs from Tatooine and their human population in SWTOR from the January 29, 2017, census.

Ord Mantell and Tython are specialized planets where avatars belonging to the Republic begin their lives, which one depending upon the class of the avatar. Each faction has four main classes, with subdivisions within each. The two classes that begin on Ord Mantell are ordinary people, having no magical powers, thus the origin point for non-Force characters. Troopers are loyal soldiers, while smugglers are

[12]swtor.wikia.com/wiki/Planets, accessed February 2017.

[13]en.wikipedia.org/wiki/Jabba_the_Hutt, accessed February 2017.

independent adventurers. Clearly law and order is an issue for each, and they respond differently, as well as having somewhat distinct technical styles of combat, forthright versus sneaky. Ord Mantell is the right training ground for them because a corrupt local government is in conflict with Separatist rebels, thus rendering conformity problematic. The physical environment is temperate and Earthlike, although some areas are inhabited by ferocious beasts.

Nearly forty thousand years before the events of SWTOR, Tython was the planet where the mystical Jedi Order was founded upon discoveries about how to harness the Power of the Force. Now, after the disastrous recent war, the Jedi have returned, in part to meditate and in part to discover lost wisdom from those past centuries, possibly hidden in the ancient ruins. Like Ord Mantell, the environment is temperate and Earthlike. Of particular interest, a village of feeble refugees belonging to the intelligent Twi'lek species struggles to re-establish their traditional religious culture, even as the vastly more powerful Jedi do the same. The two SWTOR classes that originate on Tython are both masters of the Force and carry light sabers. However, the Jedi Knights tend to leap forward and engage the enemy directly in melee combat, while the Jedi Consulars tend to stand at a distance from the enemy and hurl bolts of energy.

The large difference in the census population, 494 for Ord Mantell and 1,167 for Tython, reflects in two ways the importance of supernaturalism in the *Star Wars* mythos. First, as replicated in many other censuses I have done, the magical classes are more popular, so more players begin on Tython. Second, there are more reasons for advanced avatars to return to Tython for special missions. This may seem a very minor point in a book about computer simulation of space-related societies, but it deserves some thought. Remarkably, a few thousand people in the world today have adopted Jedi as their actual religion, in a number of small groups I have documented through research on their website forums and Facebook groups (Bainbridge 2017). The semi-reality of Jediism is also reflected in some real-world national censuses that ask the religious affiliations of their citizens. The 2001 census of England and Wales tallied 390,127 Jedi, and there were also 70,509 Jedis in Australia, 21,000 in Canada, 53,000 in New Zealand, and 14,052 in Scotland.[14] To be sure, a large but unknown fraction of these people were really secularists using Jediism as a way of protesting the influence of religion in public life, but a small core of sincere Jedis also exists.

Coruscant is the capital world for the Republic, where all Republic avatars converge at about level 10 of the 70 current levels of experience, conveyed there from Ord Mantell and Tython by a shuttle spaceflight system, before they have gained their own spaceships. The planet is one vast city, reminiscent of the one in the 1930 science fiction movie *Just Imagine* and probably derived in part from the planet Trantor in Isaac Asimov's *Foundation* series of novels. Coruscant is featured in all three films of the prequel series: *Star Wars: Episode I: The Phantom Menace* (1999), *Star Wars: Episode II: Attack of the Clones* (2002), and *Star Wars: Episode III: Revenge of the Sith* (2005). It has its own Wikipedia article, which describes it as

[14]en.wikipedia.org/wiki/Jedi_census_phenomenon, accessed June 2016.

> …a planet in the fictional *Star Wars* universe. It first appeared onscreen in the 1997 Special Edition of *Return of the Jedi*, but was first mentioned in Timothy Zahn's 1991 novel *Heir to the Empire*. A city occupying an entire planet, it was renamed Imperial Center during the reign of the Galactic Empire (as depicted in the original films) and Yuuzhan'tar during the Yuuzhan Vong invasion (as depicted in the *New Jedi Order* novel series). The denonym and adjective form of the planet name is Coruscanti. Coruscant is, at various times, the capital of the Old Republic, the Galactic Empire, the New Republic, the Yuuzhan Vong Empire and the Galactic Alliance. Not only is Coruscant central to all these governing bodies, it is the navigational center of the galaxy, given that its hyperspace coordinates are (0,0,0). Due to its location and large population, roughly 2 trillion sentients, the galaxy's main trade routes - Perlemian Trade Route, Hydian Way, Corellian Run and Corellian Trade Spine - go through Coruscant, making it the richest and most influential world in the Star Wars galaxy.[15]

The fourth environment listed in Table 8.2 is not a planet, but a fleet of spaceships belonging to the Republic, dominated by a huge, circular space station. It is the hub of transportation in many ways, beginning with the fact that the shuttle routes to the three uncontested Republic planets all meet at the fleet. In addition, each advanced avatar has a Fleet Pass that facilitates instantaneous teleportation from anywhere in the galaxy, that also places the avatar's spaceship in a dock at the station. The geometry of the space station can be explained it two ways. First, as the hub of commerce and transportation, it allows an avatar to walk between services like the bank, the player auction house, resource or equipment vendors, and skill trainers. Second, ever since the hugely influential Collier's magazine articles about spaceflight published in the early 1950s and the 1968 movie *2001: A Space Odyssey*, a standard conception for a space station is a wheel-shaped structure that rotates to create artificial gravity. However, the SWTOR space stations do not rotate, the spaceships are designed like terrestrial craft, and the entire *Star Wars* franchise ignores the laws of physical forces, in favor of the spiritual Force. The high populations for Corescant and the Fleet reflect the fact that both environments are centers of commerce and social life, to which avatars return throughout their lives.

As the lower half of Fig. 8.2 reports, the Empire has a similar structure of environments for early-stage avatars and life-long socio-economic activity, but offers three planets with rather different qualities, thus encouraging a player to create avatars in both factions, in order to experience the widest range of adventures on strange worlds. Hutta, as the name suggests, belongs to Hutt gangs, and is the point of origin for Imperial agents and bounty hunters, who are comparable to troopers and smugglers. Korriban was the ancient home of the Sith cult, that chose the Dark Side of the Force and thus opposes the Jedi. Its environment is desiccated, a little like Tatooine, but with redder sands and a network of ruined underground tombs inhabited by various monsters. There, Sith warriors and Sith inquisitors begin their stories, with skills comparable to Jedi knights and Jedi consulars. Dromund Kaas has an Imperial city in the north, and an ancient temple northwest, but is largely Earthlike jungle, being the redoubt for the temporarily defeated Empire that is readying its counterattack. The Imperial fleet is indistinguishable from the Republican fleet.

[15] en.wikipedia.org/wiki/Coruscant, accessed February 2017.

After avatars complete a number of missions on Coruscant and Dromund Kass, reaching about level 16 in experience, they are sent to one of 20 other planets, including Tatooine about level 24. They do so in nearly the same order for Republic and Empire, with the exception of Balmorra and Taris, which members of one faction visit earlier than the other. Table 8.3 lists the 20 planets, in order of increasing distance from Tatooine. Rishi is astronomically nearest, but usually not visited until level 55, thus quite distant in terms of a visitor's series of adventures. It is not only hot, humid and tropical, but is obviously a science fiction trope on stories of pirates in the Caribbean Sea. All 20 of the planets are exaggerations of real Earth locations, some more specific than others.

Scenes of Tatooine in the original *Star Wars* movie were filmed in Tunisia. The brief shots of a Rebel base at the end of it claimed to be on a jungle moon named Yavin 4, but Wookieepedia reports: "The real-world location of Yavin 4 is the ruins of an ancient Mayan city at Tikal, Guatemala."[16] Hoth is the ice planet shown at the beginning of the second movie, *The Empire Strikes Back*, while Alderaan appears in the first film only briefly, destroyed by the Death Star's laser canon in a mere practice. Most of the 20 planets have majorities of players currently operating an avatar that belongs to the Empire.

Tatooine is not the only planet where it is possible to purchase a virtual home. By the end of the research, my avatars shared strongholds on four planets in addition to that desert world. Apartments could be purchased on both of the factional capitals, Coruscant and Dromund Kaas, as well as on the shared Hutt world, Nar Shaddaa. Indeed, Table 8.3 shows a high population there, just over two thousand people, many of whom may have been holding meetings with friends. Strongholds may be owned by guilds as well as by individuals, and guilds may also purchase flagships moored at their factional fleet. A fifth stronghold was released in 2015, an extensive mountaintop temple on the planet Yavin 4, shown in Fig. 8.3.

The two figures in the foreground are standing at the edge of a cliff, looking out across the forest far below. The small one wearing a hooded cloak and with two piercing eyes is a Jawa, native to Tatooine, but now serving as a companion to Stochasta, a red-eyed, blue-skinned Chiss humanoid. Between them, in the background, is the gateway into their temple, an opening twice Stochasta's height. The Chiss are a mysterious people, allied with the Empire, but somewhat independent, as suggested by their article in an SWTOR wiki:

> As remote and secretive as the distant star system from which they emerged, the Chiss species remains a mystery to most of the galaxy. Completely removed from the Republic and the Empire, these blue-skinned humanoids evolved an advanced civilization known as the Chiss Ascendancy in the Unknown Regions of space. Despite constant political power struggles among the ruling class, the Ascendancy maintained strict controls over its dominion and the civilization prospered. This is how the Chiss developed socially and technologically in isolation for thousands of years… until they were discovered by the Sith Empire.[17]

[16] starwars.wikia.com/wiki/Yavin_4/Legends, accessed February 2017.

[17] swtor.wikia.com/wiki/Chiss, accessed February 2017.

Table 8.3 Diverse, contested planets often inhabited by intelligent species

Planet	Ecology	Region	Parsecs	Population	Empire (%)
Rishi	Hot, humid, tropical	Distant outer rim	6,032	365	59.5
Makeb	Oceanic, pillars, caves	Hutt space	10,247	406	63.3
Belsavis	Much ice, prison	Distant outer rim	12,388	576	59.4
Nar Shaddaa	A large moon, urban	Hutt space	13,347	2,016	61.3
Manaan	Oceanic, rifts	Coreward worlds	14,022	80	53.8
Corellia	Earthlike, urban	Coreward worlds	14,968	592	51.5
Hoth	Ice and snow covered	Distant outer rim	16,023	801	59.7
Quesh	Toxic chemicals	Hutt space	17,359	288	51.7
Balmorra	Plains, industrial waste	Coreward worlds	17,485	1,286	60.8
Alderaan	Earthlike, mountains	Coreward worlds	18,839	884	62.3
Darvannis	Desert	Hutt space	20,613	242	50.8
CZ-198	An exotic moon	Unknown regions	25,826	50	46.0
Odessen	Oceans, forests, mountains	Wild space	26,261	951	61.6
Voss	Plains, forests, mountains	Hutt space	26,532	725	54.5
Taris	Polluted urban, swamps	Seat of the empire	27,238	929	40.8
Yavin 4	A jungle moon	Seat of the empire	28,995	791	58.8
Ziost	Dark forests, barren tundra	Seat of the empire	31,156	157	59.2
Ilum	Arctic, ice, snow, glaciers	Unknown regions	31,774	370	65.4
Zakuul	Ocean, swamps, urban	Wild space	33,905	503	59.8
Oricon	A rocky, volcanic moon	Seat of the empire	34,696	102	54.9

Fig. 8.3 An avatar and her companion at their mountaintop temple

Stochasta is a professional bounty hunter, a standard role in the *Star Wars* universe, presented in great complexity in *Star Wars Galaxies* and one of the eight main classes of avatar and stories in SWTOR. While intelligible to today's humans, this role represents a very alien ethical system. A rich and powerful person who has been cheated or otherwise abused by some other person, often also rich and powerful, can legitimately advertise a bounty on the head of that enemy. A professional bounty hunter may then kill or capture the target and collect the bounty, without feeling at all criminal. Chiss may make especially good bounty hunters, because they are culturally and perhaps biologically remote and secretive. A bounty hunter is not emotionally violent, but shrewd and methodical.

Computer simulations are laboratories where experiments can be carried out, often by changing the value of one variable and determining the change in results. Given the strong focus on stories, and on having story-related companions, I experimented by creating a character on the planet Korriban, named Gernsback after the science fiction pioneer Hugo Gernsback, and never letting him complete any of the missions assigned by quest-giver non-player characters. Only by completing portions of the main story line could he gain companions and his own spaceship. Lacking a companion to help him, battles with monsters were more difficult, but survivable if he was careful. Lacking a spaceship, he was very limited in where he could go in the galaxy. At level 7 of the 70 levels of experience, he was able to take a shuttle to the Imperial fleet, where other shuttles could transport him to Hutta and to Dromund Kaas where he reached level 18 simply by slaughtering monsters. At that point, the experience-gain algorithm would have allowed him to gain more levels in the database, at very great effort, but in battles the algorithms for Dromund Kaas treated 18 at the maximum. He was able to teleport to a fancy home on Nar Shaddaa, but not able to leave it to visit the planet itself, thus unable to explore the universe beyond the Empire's own territories.

8.4 Simulated Aliens

One reason for taking avatars in all eight of the playable categories up to the original experience level cap of 50 was so that each of them could complete the story chapters required to acquire all five of the class-specific available companions, for a total of 40, half of whom represented humanoid alien species. Extensive information about all of them was available both at Wookieepedia and the SWTOR wiki, but directly interacting with them provided a good basis for understanding their written descriptions. There were two Twi'lek companions, a female named Vette who served Sith warriors like Sociopathy, and the other a male named Zenith who belonged to one class of Jedi, thus allowing both the Empire and the Republic to have a Twi'lek companion, and providing both female and male representatives of this interesting species.

Wikipedia explains, "Twi'leks are humanoids easily distinguished by the twin tenticular appendages that protrude from the back of the head. These prehensile appendages, known as 'head-tails', 'lekku' or 'tchun-tchin,' are advanced organs used for communication and cognitive functions... Young female Twi'leks are graceful and beautiful, and are highly sought as dancers or slaves, often sold by their own family members for profit."[18] The reference to female dancers derives from the early scenes of the third movie in the series, *Return of the Jedi*, in which Oola is a female slave dancer who belongs to Jabba the Hutt.[19]

Eighteen other intelligent alien species are represented among the companions by either a male or a female, and clearly are all two-gender life forms. They vary in how much they resemble humans, but all are bipeds with two arms and hands, a head on a neck above shoulders, and wearing clothing or armor that might fit a person. Most easily breathe the same air that we do, and have a spoken language. Table 8.4 lists these species, with information adapted from the SWTOR wiki, Wookieepedia, and from observing the companions. Players may also create avatars belonging to six of these alien species: Cathar, Rattataki, Sith, Togruta, Twi'lek and Zabrak belonging to either of two subspecies.

Despite detail variations, all these aliens have human form, and the roles they play are comparable to humans, with minor cultural differences. The one familiar alien in the early *Star Wars* movies who was most different from Earthlings in physical form was Jabba the Hutt, so fat as to be immobile and apparently lacking legs, more like a giant slug with arms than a vertebrate. Figure 8.4 shows two Hutts, who are

---

[18]en.wikipedia.org/wiki/List_of_Star_Wars_species_(P%E2%80%93T)#Twi.27lek, accessed February 2017.

[19]starwars.wikia.com/wiki/Oola, accessed February 2017.

Table 8.4 The intelligent alien species who can serve as companions

Species	Physical description	Cultural background
Cathar	Feline… normally covered in fur, although variations in the genetic baseline have expressed themselves as at least two distinct subspecies	Known for their loyalty, passion, and temper, great warriors and dedicated, efficient predators, high moral values, learned from both family and society
Chagrian	Amphibious… taller than a human with blue skin ranging in tone from light blue to indigo… two fleshy growths protruding from the sides of their heads; males have horns	A well-ordered society, abstaining from conflict but prepared to apply their rigorous attention to detail if forced into conflict
Dashade	Hulking, hairless giants of various colors, a large mouth fixed in a fanged frown, and claws on both hands and feet	Immune to the Force and rather fierce, they make good warriors, but their planet was destroyed by a supernova, and few survive
Devaronian	Very human in form, possess brightly colored skin, often red, with sharp teeth, sharp chin and nose, and the males have long cranial horns reminiscent of devils	Their homeworld has a matriarchal government, because gender dimorphism makes females more settled and serious minded, while males wander the stars
Gand	Insect-like with spherical heads and somewhat large eyes, yet human in form; they require a special breathing apparatus for ammonia rather than oxygen	A hive mind species, and most members do not even have names, but some are "findsmen," seeking a worthy goal, through perception and analysis, thus becoming individuals
Houk	Massive, muscular, physically strong and resilient, with ridged or rippling skin and huge chins that rest on their chests	Stereotyped as mindless brutes, they are in fact violent, non-intellectual, and unrestrained by laws or contracts, but not lacking in cunning
Jawa	Diminutive, rodent-like but almost completely obscured by a hooded cloak, except for bright, featureless eyes	Currently functioning as scavengers and merchants with a reputation for swindling, it is hard to know what their traditional culture was like
Kaleesh	Very human in appearance except for huge clawed hands, and a face with tusks and a toothed proboscis, although part of their faces may be masks	The culture valued honor, action more than words, and the triumph of strength over weakness in war
Mon Calamari	Although mostly human in shape, they are amphibians with frog-like faces, and their protruding eyes on the sides of their heads may move separately	Creative and intelligent, able to be competent inventors and artists, or criminals and gangsters
Rattataki	Hardly distinguishable from humans, they are pale-skinned and bald	Possibly an ancient offshoot of humanity; life on a harsh planet made them violent despite frail bodies, clever and dissatisfied

(continued)

Table 8.4 (continued)

Species	Physical description	Cultural background
Sarkhai	Possibly a different race of humans rather than an independent species, their biological distinctiveness is unknown	Of uncertain origins, living on a heavily forested planet they developed a culture of face-painting to frighten enemies
Sith	Red-skinned and often with strands of flesh hanging from the face, they are often hybrids rather than purebloods	A primitive culture recruited to a Dark Jedi religious movement exploiting the Force with passionate emotion
Talz	Covered with long, thin feathers, they are humanoid in form but have four eyes, two small ones and two big ones	A clannish species with an essential sense of honor that may become bloodlust
Togruta	Multicolored skin pigmentation and large, hollow horns that extend slightly up, then far down onto the chest, and that sense ultrasonic waves	Tightly knit culture valuing tribal unity with affinity for teamwork and togetherness, loyal and dependable
Trandoshan	Reptilian with alligator-like jaws, tough scaled skin, infrared vision and the ability to regenerate lost limbs	Culture revolves around combat and an all-knowing goddess called Scorekeeper who grants gamelike points for honorable kills
Weequay	Very humanoid but distinguishable by their coarse, leathery skin and bald heads	A laconic people deeply rooted in ancient traditions; their culture is structured around reverence of a pantheon of gods, notably Quay, the God of the Moon
Wookiee	Tall, ursine but with flat faces, covered with thick brown hair, able to understand language but able to make only limited speech sounds	Forest dwellers valuing loyalty and courage, a custom of complimenting fellows by grooming their hair
Zabrak	Distinctive facial horns, whose number and pattern distinguishes the subspecies; resistant to pain and having a second heart	Proud, even arrogant; expressing independence, confidence, and fierce determination

non-player characters, being visited by my Sith avatar, Sociopathy, and his Twi'lek companion, Vette.

One of Vette's wiki pages outlines her chaotic history before entering Sociopathy's service: "Few people have seen as much of the galaxy as Vette and few have had as little control of their destiny. Vette was separated from her family at an early age and sold to a series of minor crime lords. When legendary pirate lord Nok Drayen utterly destroyed her latest owner's holdings, Vette and the other slaves were given their choice of freedom or joining up with Nok. Vette became a pirate, travelling the known worlds and learning to get in and out of places she wasn't allowed."[20] Twi'leks tend to be victims, often slaves, but adapting to circumstances and seeking autonomy, just as human beings might.

[20]swtor.wikia.com/wiki/Vette, accessed December 2017.

Fig. 8.4 Two hutts, a pureblood Sith and a Twi'lek

8.5 Manufacturing Work

We can speculate whether the crafting skills in *Tabula Rasa* would have become more important had Arieki and Foreas not been destroyed in 2009. It is even possible that the rumored third planet might have been like Tatooine in *Star Wars Galaxies*, permitting fully simulated colonization, given that Garriott's earlier success *Ultima Online* gave great emphasis to the building of houses, furniture, and all kinds of equipment. While not placing crafting at the center of the experience, SWTOR did from its very beginning offer a complex system for creating equipment and other small resources, so it can be a good if limited example here. A defining feature of the systems is that in several ways it sought to be social, requiring exchanges between players.

In the early days, much assembly took place at fixed manufacturing facilities, with the main avatar doing much of the work. Figure 8.5 shows a trooper standing beside a cybertech workstation in the space station of the Alliance Fleet, on May 13, 2012, with an astronomical vista seen outside the window. He is making a synaptic awareness chip, from these raw materials: 4 units of fibermesh and 2 of bondite that he had obtained by scavenging, plus two bottles of brazing flux bought from a nearby vendor. While the trooper must stand near the workstation, he plans to have another one of his companions, the alien Yuun of the Gand species, manage the production, because he has an increased slicing efficiency of +10. Standing beside him at the

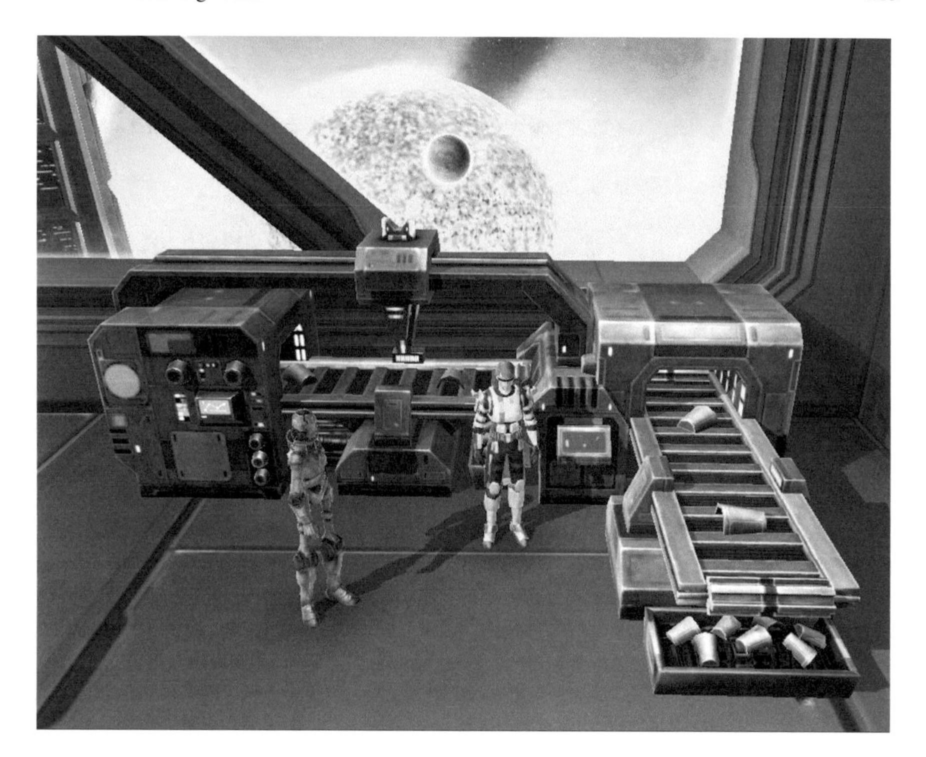

Fig. 8.5 An Alliance trooper preparing to produce a chip at a cybertech workstation

moment is his rather useless protocol droid, C2-N2, who mainly provides virtual politeness.

By early 2017, the main avatar had been liberated from manufacturing work, and companions could do it alone, hypothetically laboring at a workstation, but not one that was ever shown. The main avatar did not need to have the raw materials in the inventory available on travel, but in the abstract central storage space alternately presented as the spaceship's hold or a bank vault. Companions no longer had capabilities that allowed them to do better at one skill versus another, but a companion with a higher influence rating could do any kind of work faster. As in the early days of SWTOR, the avatar and team of companions was limited to three professions, in three categories, only one of which could be a crafting skill. As summarized on a *Star Wars* wiki, these are:

> Gathering skills:
>
> Archaeology (2,886): to seek out imbued items like lightsaber crystals and ancient artifacts
>
> Bioanalysis (1,773): collecting genetic material from creatures and plants
>
> Scavenging (3,090): recovering useful materials and parts from old or damaged technology
>
> Slicing (1,430): accessing secured computer systems and lockboxes to acquire valuable data and rare schematics
>
> Mission skills:

Diplomacy (1,068): conducting and managing negotiations

Investigation (697): examining evidence and following clues to discover valuable secrets

Treasure Hunting (1,291): to track down and recover valuable items by investigating a series of clues

Underworld Trading (1,871): the trading of illegal goods and services

Crafting skills:

Armormech (816): to work with hard metals and electronic shielding to construct all types of personal armor

Armstech (401): constructing blasters, blaster rifles and upgrades

Artifice (117): constructing Jedi and Sith artifacts

Biochem (195): engineering of performance-enhancing chemical serums and biological implants

Cybertech (489): to construct gadgets and components for droids and high-tech armors

Synthweaving (580): creating lighter outfits and armors that are imbued with Force qualities[21]

The numbers after the names of the skills come from a comprehensive survey of the sales offered on the Galactic Trade Network, which is a player auction system built into the simulation, permitting buying and selling even across the two factions. These auctions are limited to a given Internet server, so eight avatars were used to check the auction systems of the North American servers, on the last Saturday of January 2017. For the gathering and missions skills, the numbers represent how many different sales offers for materials there were at one point in time, from each of these eight skills, thus perhaps a rough measure of their relative popularity. This is much smaller than the number of units of materials, because they are usually sold in batches. The numbers for the six crafting skills represent schematics for sale that could be used by each of the crafting skills, and each sale is of just one schematic. Using a schematic confers a permanent skill on an avatar, so only one is needed.

After having become rather familiar with most of these skills, it seemed worthwhile to set some clear manufacturing goals, both to provide highly relevant examples for a book on computer simulation of social behavior related to spaceflight, and to verify that most of the technical details had been understood. Therefore, I undertook two experiments, concerning how to build the vehicles an explorer needed to ride across the surface of any of the planets. First, one level-50 character named Stochasta who initially had no crew skills would practice those necessary to make three such vehicles, at three levels of difficulty. Second, four other avatars that had reached or surpassed level 50 would cooperate to create four more advanced vehicles that harmonized with the four different gathering skills.

The *Star Wars* mythos contains many kinds of personal vehicles that somehow levitate above the surface of a planet, like the one shown in Fig. 8.2, including some called *hoverbikes*. These can be made in SWTOR by any avatar with the cybertech skill, so Stochasta selected that as her first and only crafting ability. The raw materials can be collected by *scavenging* among the gathering skills, and *underworld trading*

[21]swtor.gamepedia.com/Crew_Skill, accessed February 2017.

among the mission skills, so she also learned those, thus winding up with one skill from each of the three categories.

Gathering skills can be practiced in two ways, either by finding nodes of raw materials along the way while adventuring across planets, or by sending companions off to do the job at some significant cost in virtual money, but no waste of the avatar's time, and no computer display of them laboring away. In contrast, underworld trading can be done only by companions. Therefore Stochasta did the scavenging herself, which required returning to several of the planets she had already visited on her path to experience level 50, and often battling enemies who happened to be standing near a resource node. In the case of scavenging, an especially efficient source of materials is the wreckage of an enemy robot after it was defeated in battle by the avatar. Therefore, Stochasta especially sought out locations where high-level enemy droids abounded.

The first vehicle Stochasta was able to build was a Custom-Built Hoverbike, that required practicing cybertech enough to reach level 150 on the skill ladder that had a maximum of 600 at the time. At skill level 80, however, she could start making one of its components, bronzium cyber assemblies, 16 of which would be required. Each of these required 2 units of plastoid and 2 units of bronzium, both of which she obtained through scavenging, plus 2 bottles of conductive flux purchased from a non-player vendor character. To complete the hoverbike, she also needed 22 units of xenolite, that she obtained through underworld trading after reaching level 160 in that mission skill. A more advanced Custom-Built Speeder Bike could be constructed with a similar set of ingredients once her cybertech had reached skill level 300. Her third and final manufacturing goal was a Hotrigged Speeder Bike that required cybertech skill level 400.

Here she encountered difficulty. She could not make anything with her cybertech skills, without first having obtained a schematic, essentially the plans for the device, and learning it. Schematics for ordinary cyber assembly components could be bought rather cheaply from a cybertech trainer on the fleet space station, but they did not offer bike schematics. Apparently, these schematics were obtained at random by somebody practicing the computer-focused gathering skill called *slicing*, which is the *Star Wars* synonym for *hacking*. At rather reasonable prices she had been able to buy the first two bike schematics from the avatars of other players through the Galactic Trade Network, which is a player auction system built into the simulation, permitting buying and selling even across the two factions. The schematic for Custom-Built Hoverbike had cost her 10,000 credits from an avatar named Cylusion, and remarkably the schematic for the more advanced Custom-Built Speeder Bike cost only 1,553 from an avatar named Xaderthree. The third schematic, for a Hotrigged Speeder Bike, was available through the auction system from an avatar named Neonyx, but for the exorbitant price of 500,000 credits, when Stochasta possessed only 546,988. She would need a lot of that money to pay for raw materials. So she passed on that purchase and hoped that a schematic would soon be offered at a lower price. But for several days none were available, regardless of cost.

Frustrated, Stochasta tried a different approach. While using her gathering and mission skills, she had occasionally obtained by chance a schematic for a different profession, in particular armormech and synthweaving. Online information sources suggested that cybertech schematics could be obtained occasionally through slicing. She built her cybertech skill up to 416, so she could make the third bike, and collected all the necessary materials, getting scavenging to 423 and underworld trading to 420. She then abandoned her scavenging profession, and learned slicing, deciding she would develop skill not by personally hacking the computers on several planets, equivalent to the way she had practice scavenging, but more simply by sending out companions on a few hundred slicing missions. According to the online rumors, the particular schematic would begin to be available after reaching level 350 in slicing, in production level 6 of the 10 levels of missions, which tended to cost around 2,000 credits each to run.

The exact algorithm is unknown, but presumably several factors combine to determine the small probability that the particular schematic would be added as a bonus to the mission. The missions themselves have four levels of cost and probable return: rich yield, bountiful yield, abundant yield, and moderate yield, as well in the case of slicing some so-called lockbox missions that reward with money rather than raw materials. After a while, Stochasta deduced that the schematics were likely to be obtained through lockbox missions. It may be that skill level in slicing also matters. Upon reaching skill level 400 in slicing, she got access to production level 7 missions, but the desired schematic was said to be available only at level 6, so she faced the question of whether to continue to do only level 6 missions, or get her slicing up to the maximum 600 with level 10 and then go back down to level 6, hoping that would improve her chances. When her frustration reached the breaking point, she indeed invested the time and money to reach 600, but dozens of further slicing missions at production level 6 failed to produce the schematic.

Stochasta had been doing this labor on an Internet server called Jedi Covenant, while the four first-generation avatars were on one that gave more emphasis to role-playing, called The Ebon Hawk. One of the schematics for creating a Hotrigged Speeder Bike appeared for sale there, so one of the other avatars bought it for 200,000 credits, and Stochasta paid 90 cartel coins to transfer to The Ebon Hawk, lugging the required materials with her, and quickly was able to mass produce the bike she had labored so long to make. Except for the purchase of schematics, Stochasta had been able to do everything herself, with the help only of her simple artificial intelligence companions. A comparative experiment was having the four first-generation avatars collaborate to build distinctive speeders for each of them.

Vectron Speeders could be built by the companions of an avatar with skill level 500 in cybertech, using materials obtainable only through one of the four gathering skills, and the graphic depicting the speeder would be appropriate for that particular gathering skill. For example, the Vectron Speeder for an archaeologist would require an archaeologist to gather a certain number and type of artifact fragments and power crystals, and would appear to carry red, blue and green power crystals in the cargo space right behind the rider. The four first-generation avatars developed the four gathering skills up to the maximum 600, as well as cybertech and three mission skills required to collect specialized materials. It did not matter that the four avatars belonged to the two competing factions, because they could email the materials and

Table 8.5 Vectron speeders for four different professions

Speeder for:	Ingredients	Category	Source
Archaeologist	30 Enigmatic artifact fragment	Artifact fragments	Archaeology
	30 Ruusan crystal	Power crystal	Archaeology
	20 Farium	Metal	Scavenging
	20 Molytex	Compound	Scavenging
	10 Midlithe crystal	Gemstone	Treasure hunting
Bioanalyst	30 Metamorphic cell culture	Bio compound	Bioanalysis
	30 Anodyne extract	Bio sample	Bioanalysis
	20 Farium	Metal	Scavenging
	20 Molytex	Compound	Scavenging
	10 Autoimmune regulator	Medical supply	Diplomacy
Scavenger	50 Farium	Metal	Scavenging
	50 Molytex	Compound	Scavenging
	10 Doonium	Metal	Underworld trading
Slicer	10 Adaptive circuitry	Sliced tech part	Slicing
	60 Biocell memory core	Sliced tech part	Slicing
	20 Farium	Metal	Scavenging
	20 Molytex	Compound	Scavenging

the vehicles themselves between them. Table 8.5 shows the requirements for the four functionally equivalent but visually distinctive virtual products.

It is very unlikely that many SWTOR players would go through these experiments in the way just described, because they would obtain many of the necessary resources and the schematics from other players who were members of their own social guild. My earlier publications on both *Star Wars* virtual worlds emphasized the dynamics of guild membership, as will the following and concluding chapter of this book. In this pair of experiments, the five avatars served as simulations of five players.

8.6 Conclusion

The great distances between the SWTOR planets, reported in Tables 8.2 and 8.3, render somewhat realistic the portrayal of alien planets as rather Earthlike and never requiring breathing apparatus, given that many other kinds of planets must exist between them in the galaxy. Of the alien companions, only Yunn the Gand requires a breathing apparatus to survive in human-friendly atmosphere, and even he has no trouble with the pressure, merely the chemical composition. Yet the entire virtual galaxy appears designed for habitation by ordinary people, and offering adventures vicariously appealing to members of our species. It is well known that *Star Wars* was directly inspired by the *Flash Gordon* movie serials of the period 1936–1940. Yet it is worth noting that much of the ambiance of the galaxy also reflects the general tone and plot structure of the other popular movie serials of that era, which in my own experience would include the 1935 hybrid western and science fiction serial, *The Phantom Empire*, the 1937 detective serial *Dick Tracy*, and the 1942 military serial, *Don Winslow of the Navy*.[22] With some justice, one could say that *Star Wars* is not a simulation of our future in our galaxy, but a fanciful reminiscence of bygone California culture, where its creator, George Lucas, was born and raised.

The Jedi religion is rather obviously a California version of Buddhism, perhaps especially the Zen tradition, as illustrated by the fact that Luke Skywalker's destruction of the Death Star at the end of the firm film exactly follows the philosophy of Eugen Herrigel's 1953 book, *Zen in the Art of Archery* (Herrigel 1953; Shoji 2001). Good Californian that he is, Luke Skywalker however refused to follow exactly the ideology of his Zen master, Yoda, and abandoned his spiritual training to voyage across space to rescue his friends. This is how Buddhist writer, Matthew Bortolin, summarized Yoda's teachings: "The future is a dream. It's that place beyond the horizon where we believe happiness exists - once we finish with this or achieve that. If we live like this, we'll just fall from one trap into another. The Buddhist practices of mindfullness and concentration are a wakeup call to cut ourselves free from the world of dreams and fantasies and come back to the only place we can live life and experience happiness - right now (Bortolin 2015)." Whatever the virtues of such a philosophy may be, and whether it really reflects Buddhist traditions, it does not seem to encourage the development of the radical technologies and huge investments required to create a real galactic civilization.

It is worth noting that, with the exception of the engineered armies of the Clone Wars, the *Star Wars* mythos lacks significant invention of new technologies, even though the stories span thousands of years.[23] The main characters pretend to explore their galaxy, but almost every planet they visit has been inhabited for centuries if not millennia. Thus, for all the symbolism of "a galaxy far, far away," it is a simulation of social behavior that takes place outside the current society, political system, and economy, but using a galaxy primarily as a convenient stage for enacting human dramas. *Star Wars: The Old Republic* is something of a hybrid, requiring players to invest great effort to achieve goals set by one of the multiple, parallel story lines, and

[22]en.wikipedia.org/wiki/The_Phantom_Empireen.wikipedia.org/wiki/Dick_Tracy_(serial), en. wikipedia.org/wiki/Don_Winslow_of_the_Navy, accessed December 2017

[23]starwars.wikia.com/wiki/Clone_Wars, accessed December 2017.

giving them the profound choice between light and dark sides of the Force. More salient for this book, SWTOR requires players to master a very complex computer technology, that simulates multiple cultures and economies, yet does not seem to encourage them to take any action in our real world, that might advance the cause of spaceflight.

References

Bainbridge, William Sims. 2011. *The Virtual Future*. London: Springer.

Bainbridge, William Sims. 2016. *Star Worlds: Freedom Versus Control in Online Gameworlds*. Ann Arbor, Michigan: University of Michigan Press.

Bainbridge, William Sims. 2017. *Dynamic Secularization*. London: Springer.

Barrett, Robert R. 2004. How John Carter Became Flash Gordon. *Burroughs Bulletin* 60: 19–26.

Bortolin, Matthew. 2015. *The Dharma of Star Wars*, 3. Somerville, Massachusetts: Wisdom.

Ducheneaut, Nicolas, and Robert J. Moore. 2004. The Social Side of Gaming: A Study of Interaction Patterns in a Massively Multiplayer Online Game. In *Proceedings of the 2004 ACM Conference on Computer Supported Cooperative Work*, 360–369. New York: ACM.

Ducheneaut, Nicolas, Robert J. Moore, and Eric Nickell. 2007. Virtual 'Third Places:' A Case Study of Sociability in Massively Multiplayer Games. *Computer Supported Cooperative Work* 16: 129–166.

Gold, Mike. 1988. "Flash Gordon." *Flash Gordon* 1: 25.

Herrigel, Eugen. 1953. *Zen in the Art of Archery*. New York: Pantheon.

Lowell, Percival. 1906. *Mars and its Canals*. New York: Macmillan.

Lowell, Percival. 1908. *Mars as the Abode of Life*. New York: Macmillan.

Rinzer, J.W. 2007. *The Making of Star Wars*. New York: Ballantine.

Searle, Michael. 2011. *Star Wars: The Old Republic Explorer's Guide*, 272. Roseville, California: Prima Games.

Shoji, Yamada. 2001. The Myth of Zen in the Art of Archery. *Japanese Journal of Religious Studies* 28: 1–25.

Chapter 9
Social Life on Distant Alien Worlds

The Earth is absent from *Star Wars*, because its fantasy universe disguises mundane desires in surreal clothing. The Earth is absent from the three virtual worlds explored here, because future humans have risen above our terrestrial origins. Both *Anarchy Online* and *Entropia Universe* depict in great detail the colonization of single planets so far from Earth that communications are limited with the home planet, but not completely broken. *Mass Effect: Andromeda* goes further, exploring several realistic worlds in a star cluster associated with the far more distant Andromeda galaxy. All three can inspire people with the desire to accomplish real colonization, if much nearer on Moon or Mars, but they also raise questions about how colonies can be motivated and organized. Most crucially, how will future colonists retain a connection to Earth, and how can we here today feel connected to them?

9.1 A Mature Extraterrestrial Colony

As England discovered in 1776, when colonists in North America declared their independence, it can be difficult to exert political power at great distances. The implications of that principle can be explored through computer simulations of the colonization of the planet of a distant solar system. Within the genre of massively multiplayer online games, *Anarchy Online* counts as the pioneer in several senses, although many commentators fail to notice the considerable social theory it contains, and thus give it somewhat less credit than it deserves, even as they praise it (Burn and Carr 2003). For example, Wikipedia reports:

> *Anarchy Online* is a massively multiplayer online role-playing game (MMORPG) published and developed by Norwegian video game development company Funcom. Released in the summer of 2001, the game was the first in the genre to include a science-fiction setting, dynamic quests, instancing, free trials, and in-game advertising. The game's ongoing story-line revolves around the fictional desert planet "Rubi-Ka", the source of a valuable mineral known as "Notum". Players assume the role of a new colonist to Rubi-Ka. With no specific objective to win *Anarchy Online*, the player advances the game through the improvement of

W. S. Bainbridge, *Computer Simulations of Space Societies*, Space and Society,
https://doi.org/10.1007/978-3-319-90560-0_9

a character's skills over time. After more than 15 years, *Anarchy Online* has become one of the oldest surviving games in the genre.[1]

Rubi-Ka in *Anarchy Online* is a planet of a distant solar system that is rather far along in the colonization process, having reached the point at which the civilization has disintegrated, with the significant second-order consequence that some of the technology has gone wild and presents a wide range of local dangers, even as some geographic areas possess relatively stable social organization. Much of the planet is dominated by one of two factions, the Omni-Tek Corporation or Rebel Clans, with some colonists remaining Neutral, as they struggle to survive in an environment that remains largely beyond human control. Over the years of its existence, *Anarchy Online* has not only added areas and activities in a number of expansions, but has adjusted the experience of new players somewhat, to encourage joining one of the two factions early in an avatar's history. When I sent my first avatar to Rubi-Ka in 2008, he began in a Neutral area and joined Omni-Tek only after exploring extensively, while another avatar added in 2017 began at a newer location and joined the Clans almost immediately. Thus, an experiential simulation like *Anarchy Online* may mirror historical processes in the real world, in this case an increasing bipolarization of the inhabitants of Rubi-Ka.

One of the understandable but unfortunate evolutions in social science, beginning perhaps after the Second World War, was a loss of interest in theories of the rise and fall of civilizations. Given the trauma of both world wars, we can sympathize with a growing reluctance by intellectuals to contemplate the possibility that modern civilization was on the verge of collapse. For centuries, scholars had contemplated the fall of the Roman Empire, and the seven volumes of Edward Gibbon's analysis, originally published between 1776 and 1788, had offered cause for both pessimism and optimism about the modern world. He had considered whether Christianity had caused the fall, and concluded that it may have played a role, but also helped moderate the catastrophe as the invading barbarians were converted to this gentle religion. The chief cause, in his opinion, was that Rome had over-extended itself, struggling to control too wide a territory given the economic, social, and technological conditions of the time (Gibbon 1896). Today, we can read optimism into this analysis, because long-distance transportation and communication are no longer limited to boats in the Mediterranean Sea and horse-drawn carts on the famous but primitive Roman roads.

In the first half of the twentieth century, largely in reaction to the twin traumas of the world wars, a number of social theorists and historians contemplated the possibility that all civilizations were doomed to die, sooner or later. Oswald Spengler argued that each civilization is built upon a distinctive set of ideas—*boundless space* in the case of current Western Civilization—but will lose faith as it ages, ending in chaotic disintegration (Spengler 1926). Pitirim Sorokin agreed to some extent, but believed that civilizations could revive, naturally going through roughly thousand-years *cycles of rise and fall*, if foreign conquerors were unable to erase the indigenous culture, as in the cases of the three cycles of rise and fall experienced by both ancient China and Egypt (Sorokin 1937). Arnold Toynbee contemplated the histories of

[1]https://en.wikipedia.org/wiki/Anarchy_Online, accessed April 2017.

clearly delineated civilizations in terms of a *challenge and response theory*, placing the historical burden upon the elite class, who might fail to deal effectively with new problems, thus allowing their social system to collapse (Toynbee 1947–1957).

Probably these theories lost currency, especially in the 1960s, for three reasons. First, such ideas tended to be classified as politically conservative or right-wing, and thus dissonant with the values of many younger social scientists. Second, they offered little advice on how a technologically progressive society could prevent collapse, other than continuing to promote innovation which was happening even in the absence of these theories. Third, it was not obvious how such theories could be tested, given that they seemed to be no practical way to run experiments on statistically significant numbers of civilizations. Subsequently, world-modeling simulations like the ones reported in *Limits to Growth* could at least partially deal with the problem of research methodology, by assessing whether some plausible set of axioms about small-scale human social behavior would predict the rise and fall of large-scale social and cultural structures.

The backstory of *Anarchy Online*, the historical lore that explains current conditions, might seem more congruent with Marxism, specifically with the neo-Marxist *world systems theory* (Chirot 1977). The Omni-Tek corporation had colonized Rubi-Ka in order to extract the valuable *notum*, but most colonists were oppressed workers rather than stockholders. After much suffering, the workers rebelled, but they did so not by setting up a socialist alternative, but by trying to return to the pre-capitalist forms of society, even tribalism, suggested by their faction's name, the Clans. Thus, the workers were *primitive rebels* who failed fully to understand the progressive solutions to their problem (Hobsbawm 1959; Wolf 1969). Here, Spengler, Sorokin and Toynbee might have objected, had they still been alive, that really the Clans were following the correct course, returning to the social forms from which a durable new civilization might arise. We need not resolve this hypothetical debate, but note that computer simulations of large-scale historical processes are feasible, in virtual worlds that endure for an adequate period of time.

As we shall see in some detail shortly, *Anarchy Online* presents players with the opportunity to select among many skill classes and four biological natures when creating avatars. On November 1, 2008, I created an engineer avatar, because I wanted to learn about how operation of robot assistants was programmed, and engineers were specialists in robotics. He belonged to the Solitus *breed*, the one of four genetically altered types of human beings most similar to ordinary people and less specialized than the other three. I learned how to cope with indigenous lifeforms by battling them on a beach, with the help of his Engineer Automaton I robot.

On November 7, he settled temporarily in Neutral territory, undecided whether to join Omni-Tek or the Clans, and began exploring the territory between Newland City and Borealis. Beginning December 16, he expanded his explorations to include the neutral desert town Hope and the Clan cultural city, Athen. On December 20, he visited the Clan capital Tir, entering the nightclub where automated female kickboxers perform, getting into trouble with the local police, then teleported to the remote Clan towns Bliss and Camelot. Having toured all the Neutral and Clan settlements, he applied to change his affiliation to the Omni-Tek Corporation.

Fig. 9.1 Advancing technology in Rubi-Ka Caverns

Anarchy Online simulates extensive technological enhancement of human beings. On November 11, my avatar had experienced his first brain and eye implants, increasing his abilities through bio-nanotechnology in a surgery clinic. On December 23, in the Omni-Tek metropolis, he received nano-bio ear implants. By that point he had implants in his brain, eyes, right arm, both wrists, both hands, lungs, and waist, leaving only his left arm, legs and feet unimproved. He then invested effort developing his biotech skills, producing valuable blood plasma, looked for an apartment where he could settle down, and experienced the second-order virtual reality game called *Holo World*, a game inside a game.

Figure 9.1 well communicates the human-technology relationships within *Anarchy Online*. In the foreground, my avatar and his robot stand, facing away from us, battling an enemy in one of the many underground tunnels that often serve as instances for solo players and small teams in this simulation. Their opponent is a Techrejector, a non-player character who represents something like Luddism, not merely rejecting technological advance, but willing to resort to violence against the March of progress (Thomis 1972).

In early 2017 I briefly ran a new avatar, in order to make sure I understood the current nature of *Anarchy Online*, as the background for analyzing the quantitative data to be introduced momentarily. Figure 9.2 shows my avatar in the center, standing next to another player's avatar, in the city currently called Old Athen, with the interface's map of the wider area open on the left, as I compared with older online maps. While the details cannot be seen in this publication, the map provides players

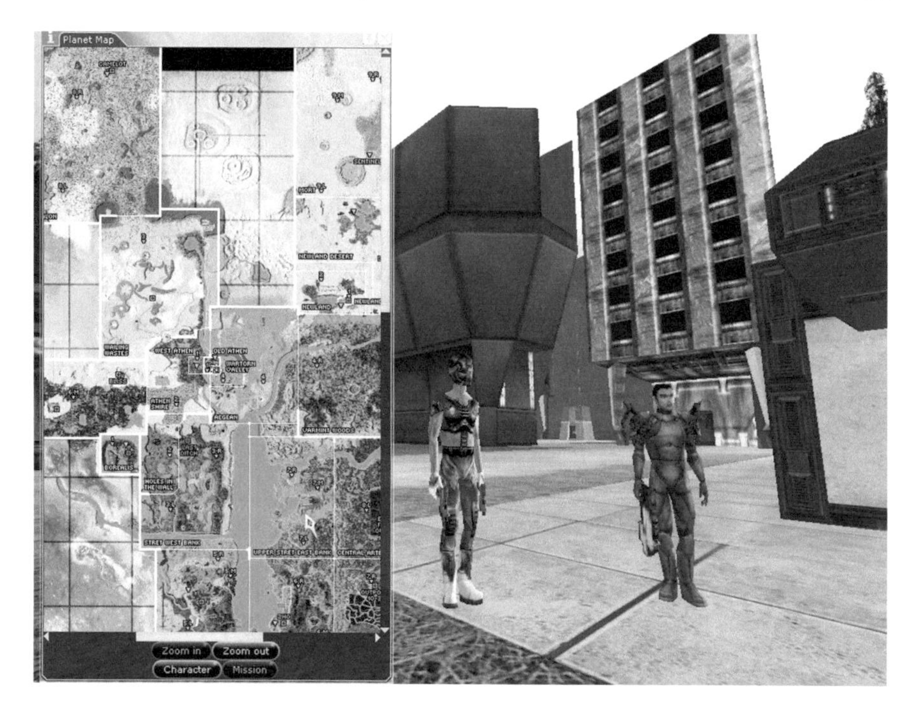

Fig. 9.2 Time and space illustrated in Old Athen on the Planet Rubi-Ka

a very useful geographic orientation. Old Athen is at the exact center of this map, and New Athen, which the avatar had just walked through, is a short distance west across a river spanned by a bridge. Old Athen is the current manifestation of the original Athen my first avatar had visited in 2008.[2]

Many versions of the in-game maps are currently available online, including edited versions uploaded by fans documenting changes to Rubi-Ka over time.[3] In addition, much information about social characteristics of players and avatars is also available online. On October 22, 2016, it was possible to download data concerning 58,400 avatars who belonged to active groups of players in *Anarchy Online*, from a database website called People of Rubi-Ka.[4] This is not really a random sample, because it does not include antisocial solo players who do not belong to groups, but is an appropriate sample for a study of social simulations. Of these, 48.8% belonged to the Clans, 42.6% belonged to Omni-Tek, and the remaining 8.5% were Neutral in the factional conflict. Table 9.1 examines their distribution across the 14 avatar

---

[2]https://anarchyonline.wikia.com/wiki/Old_Athen, accessed April 2017.

[3]https://aofroobs.com/forum/viewtopic.php?f=20&t=10481, creativestudent.com/ao/, www.gatecentral.com/ao/doc/rubika_11.php, accessed April 2017.

[4]https://people.anarchy-online.com, accessed April 2017.

classes, called *professions*, along with brief descriptions edited from the *Anarchy Online* wiki.[5]

The most popular of these professions, enforcer, is also the most physically violent, and another violent class, the soldier, is also fairly popular. The second most popular class, the fixer, may be popular because hacking computers is a fundamental activity in a computer game that simulated it. Doctors may be popular because they represent the standard MMO healer class and thus are good partners for the melee-fighter enforcers. Least popular, the shade, seems least well defined. The very large number of professions was designed to encourage players to operate many avatars, switching back and forth not only to play needed roles on a team, but also to experience variety. For example, an adventurer may explore widely, while a trader stays in a city and handles business for the adventurer and for other avatars of the given player.

The gender differences are quite significant, half of the doctors being female but only a tenth of the enforcers and a fifth of the soldiers. While we can analyze this gender imbalance in terms of traditional cultural stereotypes, it can also be understood in terms of the relationship between the player and *his* avatar, given that most players of early and violent online games seem to be male. We may theorize, but lack the data to prove, that a male player identifies directly with his male *avatar*, and prioritizes battling monsters through that personal representative. When a male player creates a female *character*, he may view her more as a friend than as an extension of himself, and give her a non-violent role in harmony with the principle of friendship.

Experience levels in *Anarchy Online* go as high as 220, and we see considerable variation across the professions. Traders may indeed be incidental tools of economic behavior for players, rather than avatars in which they would want to invest much time, and they have the lowest mean experience level. Highest are the least popular shades, which we can speculate are used as solo-play avatars because they do not perform a clear social role, and a player may have multiple role-defined avatars as well as one shade, thus dividing time investment according to an implicit algorithm that transforms a low social priority into a high experience level. The fact that the two most female professions reach high experience levels suggest that each one may indeed function as a simulated girlfriend for the male player.

As the many modifications experienced by my first avatar illustrate, *Anarchy Online* assumes that technological enhancement of the human body had become not only feasible but almost obligatory. Colonists were genetically bred, almost as four separate species, each with a different role in the corporate economy.[6] The Atrox were intended to do the brute labor of mining the notum, so they are huge, muscular, and frankly somewhat stupid. Nano Mages not only relied heavily upon nanotechnology, but had the intellectual brilliance required to manage it for the benefit of the corporation. I tend to think of the Opifex breed as the key links in the social network, likeable if often exploiting members of the other breeds. The members of the Solitus breed were least specialized, thus ready to perform all kinds of odd jobs. Table 9.2 reports differences across the breeds in terms of other variables.

---

[5]https://anarchyonline.wikia.com/wiki/Profession, accessed April 2017.

[6]https://anarchyonline.wikia.com/wiki/Breed, accessed April 2017.

Table 9.1 Statistics on the professions of avatars in *Anarchy Online*

Profession	Description	Percent of total	Percent female	Percent clans	Mean level
Adventurer	Study the animals, learn their ways and gain some of their abilities	6.6	34.8	49.2	125.5
Agent	Focus on concealment and subterfuge skills	6.2	31.2	49.8	120.6
Bureaucrat	Brings order to the chaos	7.0	40.9	48.6	143.4
Doctor	Biotechnology specialist, healer	8.5	49.5	48.3	140.4
Enforcer	Close combat using raw power and naked rage	10.1	10.7	47.7	119.9
Engineer	Specialist in creating machinery	6.9	31.8	48.6	121.7
Fixer	Getting what people need, hacking	8.9	35.4	48.7	124.4
Keeper	Fighter that radiates valour, heroism	4.7	19.9	50.6	139.7
Martial Artist	Dishing out raw combat damage	7.4	31.4	50.7	123.9
Meta-Physicist	Control multiple materialized entities and use them in combat	7.3	38.2	47.7	123.9
Nano-Technician	Expert user of aggressive nanotechnology	7.8	39.4	49.3	119.8
Shade	A mix between a predator and a parasite	4.0	39.0	49.6	157.7
Soldier	Excellence in armed combat	8.1	19.0	48.8	135.2
Trader	The ultimate entrepreneur	6.5	37.0	47.5	101.1

Table 9.2 Statistics on the breeds of avatars in *Anarchy Online*

Breed	Description	Percent of Total	Percent Female	Percent Doctor	Percent Enforcer	Mean Level
Atrox	Enormous strength and stamina, loyal, androgynous	22.0	0.0	3.4	36.7	125.5
Nano Mage	Fragile, intelligent, rely upon nano-bot super-technology	19.0	38.7	8.7	1.7	119.9
Opifex	Agile and cunning, jovial and likeable, physically weak	23.3	39.0	2.7	1.2	125.9
Solitus	Most similar to an original human being, well rounded	35.7	44.6	15.4	4.1	133.1
Total		100.0	32.4	8.5	10.1	127.3

The most normal breed, Solitus, is also the most popular, and has the highest fraction female. The fact that absolutely none of the Atrox are female does not imply they are male, because Atrox, for all their masculine traits, lack gender altogether. More than a third of the Atrox are enforcers, given their brute strength. Nano Mages have the technical abilities to be doctors, but like the non-technical Opifex are not well suited to be violent enforcers. The high fraction female among the biologically normal Solitus may help explain their higher mean experience level.

Scholars of online games often apply a theoretical concept called the *magic circle*, a cultural-cognitive assumption that walls a virtual world off from the physical world, rendering it autonomous, following its own rules and humanly meaningful despite being irrelevant to everyday life (Huizinga 1955). Economist Edward Castronova prefers the term *synthetic world* to *virtual world*, arguing not only that online social environments are subjectively real to their inhabitants, but that they indeed have very real implications for other aspects of human experience (Castronova 2005). The magic circle in *Anarchy Online* is the huge interstellar gap separating Earth from Rubi-Ka, with anarchy being the logical result of Earth governments lacking any power they can exert at that distance. Simulations play with reality, so it is entirely appropriate that *Entropia Universe* depicts colonization of an equally distant planet, yet connects it economically with Earth.

9.2 A Dynamic Extraterrestrial Social Economy

Given that massively multiplayer online games are commercial enterprises, they connect to real life economically through the payments players make to the game companies, in addition to the time and emotion they may also invest. At the present time, three different payment models exist: (1) monthly subscriptions, (2) initial payment of the full costs, and (3) occasional but necessary purchase of in-game resources. One space-related game has go so far in this third direction to make it possible for players to earn money on another planet, and return their profits to their terrestrial wallets. Colonization of the planet Calypso in *Entropia Universe* began in 2003, following an economic model in which colonists explore, collect resources, and primarily devote their time to building a market economy rather than completing many assigned missions. As of 2017, Wikipedia described this economic system in its more mature but still rather unusual form:

> Entropia uses a micropayment business model, in which players may buy in-game currency (PED—Project Entropia Dollars) with real money that can be redeemed back into U.S. dollars at a fixed exchange rate of 10:1. This means that virtual items acquired within Entropia Universe have a real cash value, and a participant may, at any time, initiate a withdrawal of their accumulated PED back into U.S. dollars according to the fixed exchange rate, minus transaction fees; the minimum amount for a withdrawal is 1,000 PED. The Entropia Universe is a direct continuation of Project Entropia. Entropia Universe entered the Guinness World Records Book in both 2004 and 2008 for the most expensive virtual world objects ever sold. In 2009, a virtual space station, a popular destination, sold for $330,000. This was then eclipsed in November 2010 when Jon Jacobs sold a club named "Club Neverdie" for

$635,000; this property was sold in chunks, with the largest sold for $335,000. Also of note: in 2014 Planet Arkadia started selling 200,000 Arkadia Underground Deeds (AUD) valued at $5.00 USD each (50PED) making the Arkadia Underground valued at $1 million.[7]

Over the two decades in which virtual worlds have flourished on Internet, a variety of systems have been used to charge players for the service. Like *Entropia*, the non-game virtual world *Second Life* charges nothing for a user to get an avatar and explore its vast environment, most of which has been created by other users. Virtual currency called Linden Dollars can be purchased for real-world money, and then used to buy virtual objects including stylish clothing crafted by other users, and to rent small areas of virtual space or housing from other users. An essential step in crafting an object using the excellent tools built into the *Second Life* software is uploading images from one's computer, by paying Linden Dollars to Linden Labs, the creator of *Second Life*. Large areas of land can be leased from the company, such as a full region capable of being visited by as many as 100 avatars simultaneously for an initial cost of $600 and a monthly charge of $295 as of early 2017. While *Second Life* continues to advertise the possibilities of profit through many kinds of virtual business, it seems that most users who gain income invest it within the virtual world, rather than taking any profits out of it.[8]

The possibility that *Entropia* colonists might earn enough to cash profits out in terms of real-world money raises the possibility, however remote, that colonization of other planets in our real universe might be funded as an investment intended to return profits. The role-playing simulations considered in this book actually offer two models of profitability of extraterrestrial colonization. The *Entropia* model highlights what might be called *interplanetary wages*, extracting terrestrial money in return for work performed beyond the Earth. The clearest alternative is *transportation fees*, analogous to a spaceflight company charging to transport to Mars rich people who wanted to visit as tourists or live there the rest of their lives. The charge may be a one-time payment, analogous to an airline ticket, or a subscription requiring recurrent payments. *No Man's Sky* can be purchased for $59.99.[9] Currently, the best deal for a premium subscription to *Star Wars: The Old Republic* is $12.99 per month, payable every 180 days.[10]

At no cost, my avatar entered *Entropia Universe* at Port Atlantis on the Eudoria continent of planet Calypso, September 20, 2007, occasionally exploring the immediate surroundings over the following months. On February 27, I purchased 482.50 PED for $50, but planned to be very frugal spending it, because my goal was exploration rather than commerce. On March 5, I bought from a non-player vendor a Sollomate Opalo laser carbine rifle for 3.80 PED, plus ammunition, so my avatar could explore Eudoria and try his hand at hunting. This had the effect of returning the 3.80 PED to the host company for *Entropia*, MindArk in Gothenburg, Sweden. Then on May 18, I purchased a LifeScanner-I from the *Entropia* auction system so he

[7]https://en.wikipedia.org/wiki/Entropia_Universe, accessed April 2017.

[8]https://go.secondlife.com/business, accessed April 2017.

[9]https://store.steampowered.com/app/275850/, accessed April 2017.

[10]www.swtor.com/buy, accessed April 2017.

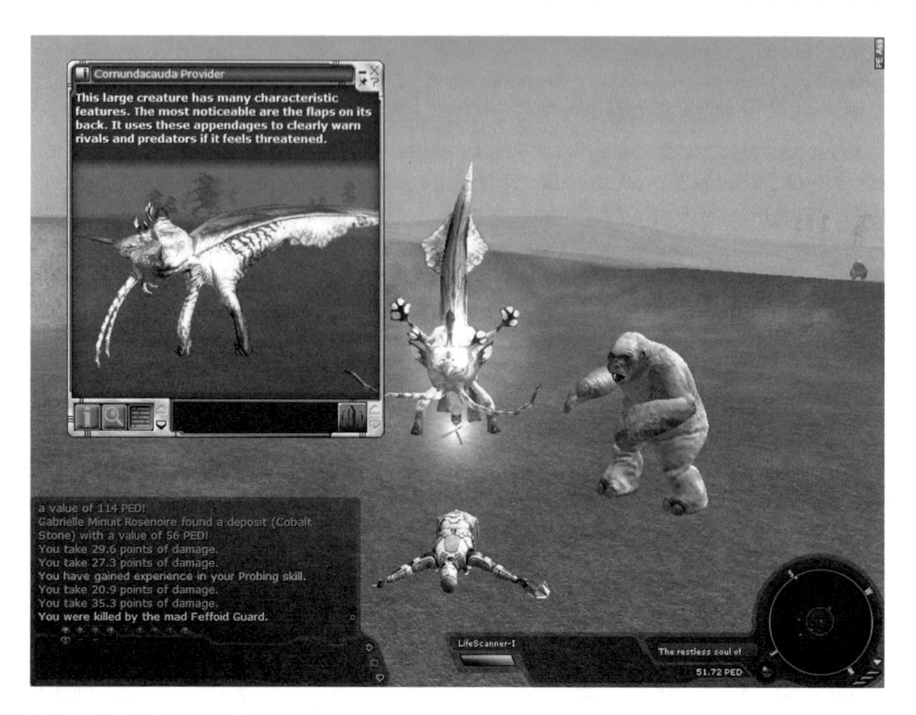

Fig. 9.3 The dangers of scientific research on extraterrestrial creatures

could begin systematic research on the animals of planet Calypso; it had been placed for sale by a player who had assumed the *Star Wars* pseudonym Dart Trader Vader, and it sold for 5.00 PED although the nominal value was 3.26 PED. On May 22, my avatar began his serious exploration of Calypso's second continent, Amethera, first checking out central areas of the privately owned Treasure Island, then ranging more widely over the following weeks.

Figure 9.3 shows both a research success and a temporary personal failure, that took place May 29, 2009. At the bottom, the avatar lies dead on the ground, both arms outstretched, although resurrection will occur afterward. He was killed by the gorilla-like creature to the right, a Feffoid Guard. The scientific gain was a scan of the creature at the center, a Cornundacauda Provider, using the LifeScanner-I. It was simply too complicated to switch quickly from the scanner to the rifle, kill the Feffoid, then resume scanning the Cornundacauda. The upper left quadrant of the image is the output from the scanner, including this text: "This large creature has many characteristic features. The most noticeable are the flaps on its back. It uses these appendages to clearly warn rivals and predators if it feels threatened."

The online datasource Entropedia has a page for this species, repeating the scanner's description and adding: "Name Meaning: 'Cornu' from the Latin for 'horn'. 'Cauda' from the Latin for 'tail'. Horn-tail. Since Cornundacauda's have a big horn and a huge tail, that kinda takes care of it. Info: A popular place to hunt Cornunda-

Fig. 9.4 Valued performance by the Calypso rescue team

cauda for the Iron Mission Chain is North Drake's on the Treasure Island (Mostly Young). For higher maturity (Old to Guardian) visit a dense spot west of Fort Zeus."[11] The Entropedia page offers several data tables, for example giving the health (hit points) for all ten varieties of Cornundacauda: Young (150), Mature (180), Old (200), Provider (220), Guardian (250), Dominant (290), Alpha (330), Old Alpha (380), Prowler (400), and Stalker (440).

The avatar's main means of travel was walking, but there also existed a fast travel teleport system with 35 stations on the Eudoria continent, and 24 on Amethera. It was necessary first to walk to each teleport station once, after which it could be visited instantly from another station. So one way of measuring my avatar's exploration accomplishments was to count how many stations had been added to the avatar's roster. On June 2, 2009, after attaining all the Eudoria teleporters and all but four of the Amethera teleporters, he became trapped in a dangerous region of southwestern Amethera. A relatively safe area was completely surrounded by dangerous beasts. Desperately hoping that there would be some way to purchase escape, I bought an additional 482.50 PED for $50 the next day, and used the in-game text chat to seek help. Figure 9.4 shows my avatar being led to safety.

The avatars of two marvelous players in the center of the picture, Coachman and Magam, belong to the altruistic Calypso Rescue Team, a group of players who would

[11] www.entropedia.info/Info.aspx?chart=Mob&name=Cornundacauda, accessed April 2017.

not charge for their lifesaving service. Coachman was well armored and led the way, battling an aggressive monster in the picture. Magam wore fancy civilian clothing and stood back a short distance, using a device to restore Coachman's health when the monster injured him. They were the Melee-Healer pair of the classic Melee-Healer-DPS *trinity* in multiplayer games, but they warned my avatar not to take the DPS (Damage Per Second) role by shooting at the monsters, because they might attack and ruin the rescue, which took a couple of hours to complete. With great care, the exploration plan to visit all 59 teleport stations was completed, then a player was paid to transport my avatar in his spaceship to Neverdie Station, a huge resort on an asteroid.

A major computational transition occurred in August 2009, migrating *Entropia* from the original graphics and database software to a more modern one. As Wikipedia reports: "With the new engine, almost everything in the game required changes, including the overall land maps. This change was retconned into the storyline as attacking robots crashing their large spaceship into the planet, changing the land."[12] On August 31, my avatar ended this phase of research by exploring Eudoria and the asteroid after this great transformation. Three subsequent research episodes were much briefer and focused on acquisition of text and statistical data about societies of players.

On February 1, 2010, I began copying and downloading the mottos, goals, and rules of player groups called *societies* with at least twenty members, from their advertisements available through a data system found in many towns. In an earlier publication, I reported: "While the official language of Entropia is English, the 183 societies included some whose languages were Chinese, Czech, Dutch, Finnish, French, German, Hungarian, Italian, Japanese, Polish, Romanian, Russian, Serbian, Spanish, and of course Swedish which is the language of the Entropia developers (Bainbridge 2011)." Again, in 2013, I returned and drew lessons about the human values that might be served by extraterrestrial colonization, for the earlier book in this series, *The Meaning and Value of Spaceflight* (Bainbridge 2015). Then, April 1–2, 2017, I checked the in-game advertisements of all 897 advertised societies that had at least 10 members, including the Calypso Rescue Team.

The advertisements use a very standard format, for example reporting that the Calypso Rescue Team was founded December 14, 2005 and currently had 74 members. Because Entropia does not require a subscription, inactive players are not automatically dropped from the membership lists, so we do not know the active membership of societies founded in previous years. The founder's name was Sweeper, while the current leader is Dinah, indicating some turn-over. Also, the "information" field of display contained what appeared to be the obituary of a respected member: "Lobo: One of the CRT's founding fathers: Gone but not forgotten may you Rest in Peace." The Team's motto was: "'You Call, We Come' Place Rescue Calls @ www.euforces.com or/join #crt in game." Indeed, the URL goes to a website for the Calypso Rescue Team that offers its history and other background information,

[12]https://en.wikipedia.org/wiki/Entropia_Universe, accessed April 2017.

including a list of the members dating from 2014.[13] The reference "/join #crt" refers to text that can be entered while in the game to join a text channel for the society. Other messages in the advertisement said: "Talk with a member if you wish to apply—we will be in touch. ALWAYS Be helpful to everyone! Have good fun!"

Societies vary in terms of how seriously they frame their official text, and not all have carefully phrased mottos, so for sake of illustration we will see only some of the clearest. The database system rates societies in terms of their popularity and separately in terms of the average achievement points scores of their members. The maximum possible size is 200 members, and the societies with the highest average achievement scores tend to be a tenth that size or even smaller. Below are the names, foundation years, and mottos of six societies with 199 or 200 members, followed by nine with the highest scores:

Earth Quakers (2003): "Gold makes the rules."

Zika Virus (2016): "The mobs are infected; it is your duty to destroy the virus."

Starfleet Command (2008): "Ad astra per aspera, every member has an equal chance."

Soldiers of Redemption (2016): "One for all, all for one."

The Entropia Order (2004): "All things to us, and a better life in this world, by helping others in the order, and get help!!! And get a good life, and a lot of things and land!!!"

Hunter Hunter (2015): "Luck always seems to be against the man who depends on it."

Art of War (2006): "Decay is beautiful!"

Dark Templar (2014): "We walk in the shadows."

Chaos Crew (2007): "Respect and Honor amongst Friends! Enjoy the Chaos."

Dark Knights (2003): "Play hard, play fair."

Warants (2003): "Live And 1337 Die."

Project X (2005): "Make -X- proud!"

Amathera Demons (2004): "Always Outnumbered - Never Outgunned."

Skyfall (2008): "When it crumbles, we will stand tall. Face it all together, let the sky fall!"

Agents of Entropy (2014): "Be Awesome."

Most of these fifteen societies were founded early in *Entropia's* history, but a third date from 2014 or after. When a society is founded, it may select from a long list of activities two that it plans to emphasize. Calypso Rescue Team chose "be helpful to newcomers" and "offer security." Table 9.3 compares the six most popular activities, for all 897 societies and separating them roughly in half by year of founding. In two cases sub-activities were listed, so for this analysis they were combined: hunting (all, animals, humans, mutants, outlaws, robots) and mining (all, energy matter, minerals). It is worth noting that the average membership size of societies founded 2003–2007 and currently having at least 10 members is 30.9, while for 2008 to April 2017 the average membership is 38.7. We may conjecture that many of the older societies are inactive, and active players who had belonged to them moved over to a more active society, which meant often a newer one.

[13]https://v2.euforces.com/?page_id=79, accessed May 2017.

Table 9.3 The activity priorities of societies (guilds) on the planet calypso

	All 897 societies and 31,179 members		456 societies with 14,092 members, 2003-2007		441 societies with 17,087 members, 2008-2017	
Activity	Societies (%)	Members (%)	Societies (%)	Members (%)	Societies (%)	Members (%)
Be helpful to newcomers	20.0	26.0	14.5	18.7	25.6	32.0
Exploring the new world	12.3	13.3	11.2	12.5	13.4	13.8
Having fun	31.8	32.0	33.3	32.1	30.2	31.9
Hunting	35.3	29.5	39.7	37.8	30.8	22.5
Mining	12.7	12.1	13.6	14.1	11.8	10.5
Personal develop-ment	17.8	20.5	18.9	25.5	16.8	16.4

The nature of the data suggest we should be cautious in our interpretation, but it is plausible that an attractive activity is not merely one advertised by many societies, but one in which the percentage of players in such societies is greater than the percentage of societies. Looking at the period 2003-2007, we see that two activities stand out by this measure. "Be helpful to newcomers" is an activity of 14.5% of societies, but 18.7% of members, a ratio of 1.3, and the ratio for "personal development" is also about 1.3. For the societies founded 2008-2017, the ratio for "be helpful to newcomers" remains about 1.3, but that for "personal development" drops to around 1.0. Notably, over the time periods the members/societies ratio for hunting drops from roughly 1.0 to 0.7. More obviously, members in societies that seek to be helpful to newcomers rose from 18.7 to 32.0%, while the corresponding fraction for hunting dropped from 37.8 to 22.5%. It is reasonable to conclude that indeed, as we might expect, societies that were especially hospitable to newcomers gained membership as a result.

9.3 Colonizing the Andromeda Galaxy

On March 21, 2017, a multi-species expedition led by humans reached the Andromeda galaxy, two and a half million light years from Earth, and two weeks later I began research on the social dynamics of colonization. Three earlier *Mass Effect* games situated in the Milky Way galaxy have impressed scholars of computer games for their social and philosophical sophistication, and the fourth in the series transcends their boundaries (Zakowski 2014; Carvalho 2015; Patterson 2015). *Mass*

Effect: Andromeda was the newest complete simulation explored for this book, older only than ROKH which is far from finished. For several reasons, *Andromeda* is the best example with which to end. Reminiscent of the strategy games considered in Chap. 5 and the *Star Wars* games, several intelligent species are depicted, all rather humanoid in form but possessing different characteristics. The colonization mission is called the Initiative, launched in the year 2185 as "a civilian, multi-species project created to send scientists, explorers, and colonists on a one-way trip to settle in the Andromeda Galaxy."[14]

Four alien civilizations of the Milky Way galaxy have formed an alliance with Humans. Asari are "known for their elegance, diplomacy, and biotic aptitude."[15] Salarians are "warm-blooded amphibians" who "possess a hyperactive metabolism; they think fast, talk fast, and move fast."[16] Turians have a "militaristic and disciplined culture" and were responsible for defeating the fourth species, the Krogan, whose rapid reproductive rate and aggressive behavior had endangered all the other intelligent species.[17] The Krogan are "large reptilian bipeds" who had "destroyed their homeworld in a nuclear war that reduced their race into primitive warring tribes," then joined the galactic community with the help of the Salarians.[18] The method by which the Turians defeated the Krogan was a biological weapon called the *genophage* "designed to severely reduce krogan numbers by 'infecting' the species with a genetic mutation" that greatly reduced the survival rate of their offspring.[19]

Clearly, the alliance of five species that launched the Andromeda Initiative is shaky at best, but when the first ship reached its destination, an unexpected disaster greatly complicated social relations. The main vessel was a huge space station called *Nexus*, on which all the passengers slept for six centuries in cryostasis pods. Upon arrival, a small number were awakened, but immediately all leaders were killed by a strange apparently natural force the survivors called the "Scourge," which also seemed to have devastated the Golden Worlds they had intended to colonize. The novel *Nexus Uprising* explains how the incompetence of surviving officials combined with the existing inter-species hostilities to cause multiple rebellions (Hough and Alexander 2017). With little chance of survival, rebels were forced to flee to nearby planets, as did the many Krogan who had been assigned to Nexus. In addition to Nexus, the first colonization fleet included four arks, each of which primarily carried colonists from one of the more cooperative species: Hyperion with Humans, Leuinia with Asari, Paarchero with Salarians, and Natanus with Turians. The player's avatar voyaged on Hyperion to a planet named *Habitat 7* that the Scourge had rendered uninhabitable.

Many of the early reviews criticized the game's computer graphics, but objectively they are quite excellent. For example, when driving the Nomad six-wheel armored vehicle across a desert, sand and stones fly realistically up in the air, and driving

---

[14]https://masseffect.wikia.com/wiki/Andromeda_Initiative, accessed May 2017.

[15]https://masseffect.wikia.com/wiki/Asari, accessed May 2017.

[16]https://masseffect.wikia.com/wiki/Salarian, accessed May 2017.

[17]https://masseffect.wikia.com/wiki/Turian, accessed May 2017.

[18]https://masseffect.wikia.com/wiki/Krogan, accessed May 2017.

[19]https://masseffect.wikia.com/wiki/Genophage, accessed May 2017.

in circles digs complex tracks on the ground. The criticisms focused on the facial expressions of artificially animated characters, such as their eye movements, but there are two reasons why they were excessively harsh. First, as computer graphics become highly realistic, they enter the uncanny valley theorized by Japanese robot pioneer Masahiro Mori, that lies conceptually between unrealistic cartoonish representations of humans and perfect realism.[20] This may apply to non-visual aspects of simulated humans as well. We may tolerate a lack of realism in computer simulations if they are simple programs based on clear theories, but as they evolve into complex reflections of real life, they may seem annoyingly unrealistic, until that moment in our technological progress that they become fully real. Second, *Mass Effect: Andromeda* is a multi-media convergence, combining role-playing action on the wide theater stage of a virtual planet, with innumerable cutscenes like segments of a movie serial, with a variety of data displays of more abstract kinds. Costing a reported $40,000,000 to produce, *Mass Effect: Andromeda* handles its logistical performance challenges well, but perfection would require an infinite budget.

The multiplayer version of *Mass Effect: Andromeda* was not suitable for exploratory research within the theme of this book, because it reveals little if anything about the meaning of planetary colonization and is almost entirely orientated toward violent combat. It is really a separate game, with no consequences for progress in the solo-player main game. It follows a standard design principle called *wave-based horde mode*, which a popular videogame news website defines as "a type of multi-player game mode in which players battle co-operatively against increasingly difficult waves of computer-controlled enemies."[21] A team consists of four players, who may be friends who schedule a couple hours together, or strangers of the same skill level assembled by the Internet-based system. A player begins by selecting one of several standard avatars, and that avatar's skill will increase from session to session, depending on how well the team does.

Each multi-player session consists of seven rounds that take place at different locations within a single environment, comparable to an area on one of the Andromeda planets. Some rounds are *objective based*—such as one team member hacking a device while the other three kill attacking enemies around it. Apparently, systematically killing all the enemies will not succeed, taking too long even if the supply of enemies were to be exhausted, so much of the strategy concerns managing a series of goals focused on particular locations, thus requiring good spatial perception as well as simply violent exercise of high-powered weaponry. If one avatar is incapacitated by wounds, a team mate may be able to revive it, and a player may also consume medi-gel, a costly medicinal salve. Apparently teams are seldom all killed, unless members are incompetent or distracted by the real world, but poor performance is

[20]https://en.wikipedia.org/wiki/Uncanny_valley, accessed December 2017.

[21]www.giantbomb.com/horde-mode/3015-3021/games/, accessed April 2017.

reflected in poor rewards at the end. The seventh round requires the team to go to an *extraction zone*, where they rendezvous with a ship that will fly them to safety.[22]

The main version of *Mass Effect: Andromeda* is a solo-player game, yet highly social in the player's perception, because the programmed behavior of non-player characters is complex, especially when enacting scenes from the sophisticated story. The player's avatar belongs to a team, each member of which has a distinct personality and speaks with the voice of a professional actor, and during the story many interactions and conversations occur with other characters. The player takes the role of one or the other of a brother-sister pair, Scott or Sara Ryder, children of Alec Rider, the Pathfinder leading the colonization of the planet called Habitat 7. A specialized wiki explains what happens:

> The planet known as Habitat 7 was one of the "golden worlds" selected by the Andromeda Initiative for early settlement. Signs pointed to a lush and biologically diverse tropical region that could easily support an outpost. With no communications from the Nexus or sister arks upon arrival in the Heleus cluster, the Hyperion approached Habitat 7 to begin survey and settlement operations as soon as possible, but even visual assessments showed that the planet was no longer viable. Habitat 7 is now a storm-wracked world with an unbreathable argon-nitrogen atmosphere. Intense magnetic activity in unknown metallic elements interacts with the storms, causing interesting but highly destructive electrical phenomena. The investigating Pathfinder team encountered hostile alien life and strange technology on the surface. Reactivating this technology caused a noticeable change in the conditions on Habitat 7. However, the planet is still unsuitable for settlement, and the resulting activation ultimately claimed the life of Alec Ryder, the human Pathfinder.[23]

The player's avatar inherits the Pathfinder role from his or her father, the responsibility to establish a colony successfully on at least one planet. All of the Golden Worlds have been ruined, each in a distinctive way, by that mysterious Scourge energy force: "It defies understanding by contemporary science, and is described as a 'dark energy cloud' or 'charged absence'. Speculation about the nature of the Scourge usually ties it to dark energy, yet the Scourge does not behave in the way that existing theories on dark energy say that it would."[24] It turns out that five of the Golden Worlds can be colonized, but this will require restoring their environments to something approximating normal. As a research plan, but identical to the goals most players would set for themselves, I determined to get each of the five planets to a *viability* measure of 100%. They are listed at the top of Table 9.4, in the order visited by Scott Ryder.

The orbital distance measures are in astronomical units (AU), which is simply the radius of the Earth's own orbit. The bottom of the table gives the approximate numbers for Venus, Earth, and Mars in our own solar system, and Earth by definition

---

[22]IGN Plays Live, "Mass Effect: Andromeda Multiplayer Livestream Reveal," www.youtube.com/watch?v=2LViubGO2H0; GameSpot, "Mass Effect Andromeda Multiplayer Livestream," www.youtube.com/watchX5sPDvWIEJA; JV2017gameplay, "MASS EFFECT ANDROMEDA: Multiplayer Ultra Rare Classes Stream! (Salarian Operator and Angara Avenger)," www.youtube.com/watch?v=YukgK_OeACc, accessed April 2017.

[23]https://masseffectandromeda.wiki.fextralife.com/Habitat+7, accessed May 2017.

[24]https://masseffect.wikia.com/wiki/Scourge, accessed May 2017.

Table 9.4 Eight planets that can be visited in the Heleus cluster

Planet	Orbital distance in AU	Orbital period years	Radius in kilometers	Pressure in atmospheres	Surface temperature, centigrade	Description after scourge damage
Colonizable planets in Andromeda:						
Eos	1.3	1.6	2,952	1.98	19	Desert world with significant resources… deadly, radiation- contaminated storms
Havarl	2.4	2.9	7,103	0.89	29	Lush jungle planet… plant life… exhibiting maladaptive, unsustainable growth patterns
Kadara	0.9	1.0	8,859	1.10	29	Mountainous… plagued by toxic water caused by an atypical amount of sulfide minerals
Voeld	14.8	57.1	8,740	1.42	−40	Going through an ice age… remains of vast ancient cities are still entombed in ice
Elaaden	32.8	44.4	6,168	0.89	52–99	Moistureless, sweltering landscape covered by a large number of cover-subsidence sinkholes
Visitable but not colonizable planets in Andromeda						
Aya	1.1	1.3	3,289	1.40	28	Small and lush tropical world.. sovereign world of the Angara
H-047c	116.7	719.1	668	0.02	−115	This formerly viable planet is now a debris field… excessive solar and cosmic radiation…
Meridian	Irregular	Irregular	1,800	1.03	N/A	A hollow sphere protecting a controlled and malleable environment
Real planets for comparison:						
Venus	0.7	0.6	6,052	90.80	462	Oppressive
Earth	1.0	1.0	6,371	1.00	15	Normal
Mars	1.5	1.9	3,390	0.01	–63	Insufficient

has a distance of 1.0 AU from the sun, the distance for Venus being less, and for Mars, more. Kadara has a orbit around its star most similar to that of the Earth, 0.9 AU, and its year is exactly the same. Should not the Kadaran year be 0.9 of an Earth year, rather than exactly 1.0? Not really, because the speed of a planet in its orbit also depends upon the mass of the star, so this implies that Kadara's star is slightly less massive than our Sun. Indeed, each of these 8 Andromeda planets is in a different solar system, with all the uncertainties that introduces into our analysis.

9.4 Normalizing Alien Environments

The radius of Kadara is 8,859 km, compared with 6,371 for the Earth. Can we predict the atmospheric pressure on Kadara, based on the planet's size? If it has the same density as Earth, we could try to calculate the relative gravitational attraction on its surface, a measure not given in the table. The ratio of the diameters, Kadara/Earth, is easy to calculate: 8859/6871 = 1.29. So, should Kadara have 1.29 times the Earth's atmospheric pressure, rather than the 1.10 we see in the table? No, because the relative masses of planets is a factor of their relative volumes, not radiuses. Figuring on the basis of volumes we get a predicted atmospheric pressure for Kadara of 1.66. Of course we do not in fact know that the densities of the two planets are the same, but the futility of this kind of assessment is illustrated by the fact that the planet Venus, with almost the same radius as Earth but slightly less, has a surface atmospheric pressure 90 times as great.

The table reports that Kadara has "an atypical amount of sulfide minerals," which rendered the water toxic, a difficult challenge for colonization. Figure 9.5 shows Scott Ryder in the center and two of his crewmates standing at the edge of one of the poisonous ponds on Kadara. The water fumes, bubbles, and would kill them if they tried to wade through it. Already the team had experienced the radioactive zones on the planet Eos, and the more subtle progressive degradation of jungle life on Havarl. One of the ironic applications of these poisonous ponds is disposal of the dead bodies of murder victims, and the team was currently on a mission to find five of them and bring evidence back to prove murders had been committed.

Many different missions are required to restore the environment on any of these planets, but fundamental is restoring function to an environmental control system that had been created by a vanished civilization long ago. Ambiguously called the Remnant by the colonists, that civilization is represented by exotic architectural ruins, wayward defensive robots that attack anyone who approaches key locations, and a system of three monoliths and a vault that constituted the environmental system on each of the five colonizable planets. The vaults were typically filled with monsters, enemies, and Remnant robots. In both monoliths and vaults, controls often needed to be turned on, for example to extended a bridge across a chasm. Key controls, usually those switching a monolith back on, required entering a complex coded command.

These crucial control devices were in the form of sudoku puzzles, either 4 by 4 or 5 by 5 squares, in which either 4 or 5 different glyphs needed to be arranged

Fig. 9.5 Exploring a sulfide hot spring on the Planet Kadara

in the proper manner. Some of the glyphs were already installed in the controls, but often two or three needed to be found where they had been hidden in the local environment. Ryder had a scanner he could use to track the locations of the hidden glyphs, but getting to them often required jumping up and over complex ruins in the local environment. Admittedly, the most efficient way to solve the puzzles was to open a web browser in parallel with the game, and look at solutions that other players had already captured and shared online. Many other missions required incredibly complex climbing sequences, using hidden controls to move objects on which the avatar could jump, and on more than one occasion I resorted to watching another player's YouTube video, following their movements step by step. In a way, this rendered a solo game social. These sudoku puzzles and three-dimensional mazes certainly seem like frivolous play, but they had the effect of requiring the player to think, giving this simulation considerable intellectual challenge for the players.

Ryder's two companions in Fig. 9.5 suggest a different kind of social-intellectual challenge that gives the simulation a degree of realism. The blond woman standing on the right side of the picture is Ryder's teacher, Cora Harper, with whom his relationship evolves, especially if he undertakes special missions for her personally. She is but one of six *squad members* who came with him from the Milky Way, each with a personality and special missions, any one of which might have been brought along on general missions like the one to find the dead bodies in the toxic pools. The other companion, Jaal Ama Darav on the left, is a native of the Heleus Cluster in Andromeda, a member of the Angara intelligent humanoid species:

> The angara are the only known sentient species local to the Heleus cluster. Scattered across numerous worlds, the angara are reuniting as a people. Though much of their culture and scientific knowledge has been lost, the angara continue to rebuild and resist the kett conquest of Heleus. From our cultural exchanges, we know that the angara have a unique control over electromagnetism. Specialized skin cells and organs allow them to generate and control

> electromagnetic fields artistically or unleash them in combat. Oral histories describe how ancient angara were tribal and nomadic before settling in cities. Many of them still live in large, tight-knit families and workplaces have a guild-like organizational structure.[25]

Table 9.3 lists the Angara homeworld, Aya, in sixth place. The capital city can be visited, but it is not open to colonization by immigrants from the Milky Way. In addition to Cora, Ryder's original squad consists of Liam Kosta, an Englishman of apparently African descent, and a Krogan male named Nakmor Drack, an Asari female named Pelessaria B'Sayle (or Peebee for short), and a Turian female named Vetra Nyx. The world designated as H-047c in Table 9.3 is a mere asteroid fragment of the original H-047 that had been the goal of the Turian ark, and so it features in special missions for Vetra Nyx, as well as being a fine location for mining minerals and helium-3. The Krogan will be described in the following section of this chapter, but here are the in-game descriptions of three other prominent Milky Way cultures:

> The asari come from Thessia, whose cast element zero deposits form the basis of the most powerful economy in the Milky Way. Living in small city-states, the natural asari tendency to cooperate led to the loose conglomerate known as the Asari Republics. Government operates as an "e-democracy;" policy discussions take place over the extranet and are open to all. Decisions are made by consensus, or with the advice of politically minded Matriarchs.
>
> Turian culture is founded on public service and personal accountability. The turian military is the center of their society, functioning as a public works organization, as well as an armed defense force. Citizenship is conferred once turians have completed boot camp (begun on their fifteenth birthday) and promotion through the 27 citizenship tiers is judged on individual merit. At the top are the Primarchs, who lead colonization clusters. Their inclination to military service and social trends mean few turians pursue entrepreneurial careers, making the turian economy more vulnerable than most.
>
> ...the Salarian Union is dominated by the rare salarian females known as "Dalatrasses," who negotiate the complex web of salarian politics on behalf of the regions they govern. Salarians belong to vast clans, whose interrelation and current political status must be painstakingly tracked... Salarians do not have a concept of romantic love. The fertilization of eggs is a political act, since the lineage of any resulting females has a strong impact on society, and only occurs after months or even years of negotiations.

A female Salarian named Kallo Jath is the pilot of Pathfinder Ryder's sleek exploration spaceship, the Tempest. All the alien characters have two legs, two arms, and two eyes, very humanoid in form. In the cutscenes they make facial expressions very much like those of humans. To be sure, it is much easier for the creators of a role-playing computer game to create smoothly operating aliens by slightly adjusting the exact geometry of already-existing human models, and uploading moderately strange graphics on their surfaces. But in a virtual world where one's avatar develops close relations with alien individuals, it helps if their kinesics and expressions are relatively conventional. How can you tell if aliens are happy, if they don't smile?

That raises the question of the extent to which virtual worlds are able to be realistic simulations of truly alien phenomena. In a blog illustrated by a screenshot of *Mass Effect: Andromeda*, Eliot Lefebvre observed: "We are constantly re-indexing our experiences so each new thing we're exposed to becomes part of the existing mental

[25] https://masseffectandromeda.gamepedia.com/Angara, accessed May 2017.

Table 9.5 Three of the 35 solar systems in the Heleus cluster

Planet	Orbital distance in AUs	Orbital time in years	d^3/t^2	Radius in kilometers	Pressure in atmospheres	Surface temp. C°	Description
Kindrax solar system:							
H-073	0.7	0.6	0.95	6282	44.86	523	Constant volcanic activity
Mendradym	1.1	1.2	0.92	7911	0.14	44	Ravaging electrical storms
Tunharaset	1.9	2.7	0.94	4822	0.03	−39	Warping effects of.. Scourge
H-202	2.8	4.6	1.04	9828	0.00	−82	Surface… of tin and boron
Civki solar system:							
H-065	1.4	1.3	1.62	5678	0.04	195	Scorching desert
Rakaelmo	3.2	4.5	1.62	4731	0.68	42	Massive storm systems
H-309	4.5	7.2	1.77	68483	1.35	?	Gas giant
H-110a	13.0	36.2	1.68	6219	0.32	−117	Satellite of gas giant H-110
Skeldah solar system:							
Bleeding Ruby	0.1	0.1	0.10	9912	101.38	361	Tectonic instability
Kotkoborra	0.3	0.3	0.30	6634	53.07	126	Ongoing volcanism
Norgraqua	0.9	1.6	0.28	4797	1.68	−117	Windy, frigid desert
H-061	3.5	89.3	0.01	6669	0.02	−207	Orbital velocity slows
H-329	4.4	16.9	0.30	4984	0.22	−241	Frozen nitrogen oceans

landscape we occupy. So sure, the *first* time you go to space, it's all about gasping in wonder. The thousandth time, though, you find yourself going into another galaxy in cryo-stasis because now space travel is *boring* and *everyone* goes to space."[26] Yet if the extraterrestrial environment is realistic, then explorers may continue to discover interesting phenomena if they employ the tools of scientific research.

Having noted the fact that the orbital mechanics of planets might possibly be realistic, I collected the data for three solar systems that a mission arc sent Tempest through, without landing. A classic astronomical formula, called Kepler's 3rd Law, relates the two orbital variables: "The ratio of the squares of the revolutionary periods for two planets is equal to the ratio of the cubes of their semimajor axes."[27] The *revolutionary periods* are identical to what the game calls "orbital time," and the term *semimajor axes* is similar but perhaps not quite identical to "orbital distance" which might mean *radius*. A circle has a radius, while an ellipse has a semimajor axis. Thus, there may be small measurement errors, as well as rounding errors, in applying Kepler's 3rd Law to the three solar systems in Table 9.5.

[26]https://massivelyop.com/2017/04/22/wrup-a-new-and-largely-identical-frontier-edition, accessed April 2017.

[27]https://physics.unm.edu/101lab/lab3/lab3_C.html, accessed December 2017.

The formula d^3/t^2 refers to the cube of the orbital distance (d) over the square of the orbital time, (t), and is simply a re-arrangement of Kepler's formula to put the data for one planet on one side of an equal sign from the data for another. Consider the first planet in the Kindrax solar system, H-073. Its t is 0.6, and the square of t is 0.36. Its d is 0.7 and the cube of d is 0.343. Then divide d by t and the result is 0.95277... or rounding off, $d^3/t^2 = 0.95$. In our own solar system, we expect these ratios to be 1.0, but the stars of other solar systems have different masses, so our expectation for them is merely that the ratio *s* for these planets should be about the same. Indeed, d^3/t^2 s about 0.95 for the Kindrak planets, 1.68 for the Civki planets, and about 0.30 for three of the five Skeldah planets. The in-game codex explains why one of the Skeldah planets has strange numbers: "H-061's orbital velocity slows as the planet passes near tendrils of Scourge, radically extending its total orbital period in ways difficult to calculate."[28] A wiki does the same for Bleeding Ruby: "Named by an exile with a penchant for poetry, Bleeding Ruby's rings and tectonic instability are the result of a single asteroid impact that occurred less than a century ago."[29] Thus the designers of *Mass Effect: Andromeda* not only followed Kepler's laws, but offered good justifications when they departed from them.

9.5 A New Home

Having considered colonization of Rubi-Ka and Calypso that took place years ago, in the period 2001-2003, we can now compare with colonization of Elaaden in 2017. In addition to the evolution of the virtual world genre over a decade and a half, the solo-player nature of *Mass Effect: Andromeda* gives one individual a special role, that of Pathfinder to explore new worlds and provide guidance to the many other simulated characters in the computer-generated environment. Here we should focus on the ecological and astronomical dimensions of the environment, as well as the social. Figure 9.6 shows Ryder early in his exploration of Elaaden, near two pieces of his equipment, with information in the sky behind him. To the left, outside the picture, is a doorway defended by two Krogan guards, leading to a cavern city called New Tuchanka, that was named after the ruined Krogan home planet.

The background sky in Fig. 9.6 suggests that Elaaden is a planet-sized moon of a gas giant planet. Two other moons are in the sky, and views of them from space indicate that they are much smaller than Elaaden. Table 9.4 reports that Elaaden is between Earth and Venus in size, yet in a vastly larger orbit, each Elaaden year taking 44.4 Earth years. The surface temperature varies between 52 and 99 degrees centigrade, remembering that the human body temperature is about 37 and 100 is the boiling point of water under an air pressure of 1.0 atmospheres. One wonders if the somewhat lower atmospheric pressure on Elaaden means that water can boil away, even at 99. Apparently Elaaden's star is much brighter than our sun.

---

[28] https://libraryofcodexes.com/codex=15880/H-061, accessed December 2017.

[29] https://masseffect.wikia.com/wiki/Bleeding_Ruby, accessed December 2017.

Fig. 9.6 A scene on the surface of Elaaden, a planet-sized moon

Before Ryder has achieved an improvement in the viability on Elaaden, open, sunlit spaces are lethally hot, and he must run quickly into a shadow or technologically protected area, or he will die within a very few minutes. His most important refuge is the Nomad wheeled vehicle behind him, which he employed to travel across the frozen wastes of Voeld, and here he can stay protected even for hours of burning daylight. Certain kinds of mining can be done without leaving the Nomad. It can be transported from one planet to another by spaceship, including in the Tempest, or on a planet from anywhere on the open surface to a forward station.

One of the forward stations appears in the right third of Fig. 9.6. It also provides a protecting environment for a short distance around it, despite the hot sunshine. As Ryder explored one of the five colonizable planets, at certain locations a forward site would be delivered by rocket as he approached its coordinates. After he manually switched it on, it could be used as a quick travel location. Each also offered access to equipment, such as weapons Ryder could carry on his person, and restoration of both health and ammunition. There also was a terminal for doing research and development, related to the crafting of resources that will not be analyzed here. At the end of his explorations, Ryder had activated a total of ten forward stations on Elaaden, including one at his outpost, shown in Fig. 9.7 with the Tempest spaceship in the background.

It had not been easy to get the right to build an outpost on Elaaden, because the disgruntled Krogan considered it their new home. Learning the entire backstory would require playing through the three previous *Mass Effect* games, but information contained within the game and on the web provides the outline. One of the *Mass Effect* wikis reports, “Krogan have always had a tendency to be selfish, unsympathetic, and

Fig. 9.7 The Elaaden settlement

blunt."[30] Their bodies are built for violence, for example with hard armor and back-up duplications of major organs. Another wiki provided historical background:

> The krogan homeworld boasts extreme temperatures, virulent diseases, and vicious, predatory fauna. Around 1900 BCE, the krogan discovered atomic power and promptly instigated many interplanetary wars, sending Tuchanka into a nuclear winter. With most of their industrial base destroyed, the krogan entered a new dark age and warring tribal bands dominated. Populations remained low for the next 2,000 years. First contact with the salarians made resurgence possible. Krogan brought to less hostile planets bred exponentially and returned to reconquer their home. They built vast underground shelters to shield themselves from surface radiation, which proved prescient during the Krogan Rebellions when many of them isolated themselves in a vain attempt to avoid the genophage. Convinced they could outbreed the genophage, they transmitted it into more than 90% of the sealed bunkers. Today, Tuchanka's population is sharply limited and while individual krogan are long-lived, the genophage ensures few replacements.[31]

As it happened, Ryder was not able to establish this outpost on Elaaden until the viability was already 100%, and the temperature was tolerable even in the bright sunshine away from any protection. He also had outposts on Eos, Kadara and Voeld. He could not establish one on Havarl, because the Angara had already settled there, but he had developed such a strong alliance with them that his immigrants from the Milky Way were entirely welcome there. Ryder's Elaaden outpost was quite large, containing much major equipment, several buildings with many rooms, and a few scientists or other early colonists transported from Nexus, all simulated as non-player characters.

---

[30] https://masseffectandromeda.gamepedia.com/Krogan, accessed May 2017.

[31] https://masseffect.wikia.com/wiki/Tuchanka, accessed May 2017.

9.6 Conclusion

When it comes time for actual colonization of Mars, we cannot be sure what lessons people may draw from simulations like colonization of Rubi-Ka, Calypso or Elaaden. Yet these virtual worlds have inspirational as well as educational functions today, and contemplating them raises real issues. Will extraterrestrial colonies extend the scope of terrestrial governments? Will they repay our planet's investments with economic profits? Or, will they chiefly be motivated by the dream that through technology humans can transcend their ancient limitations, and explore new frontiers rich with meaning? While academic models of space society may be few and often primitive, by combining them with sophisticated games aimed at the general public we have learned much about the challenges and insights associated with computer simulation of human society in outer space, and of the dynamics of terrestrial enthusiasm for space exploration.

It is plausible to imagine that the future of computer simulation of human social behavior will require progressively more massive computational systems, beyond Big Data and Huge Data to Cosmically Vast Data. However, the scope of social simulations is limited by the imprecise nature of the existing theories, which might lead to the very different prediction that simulations of the future will find new ways to incorporate human beings, without requiring highly advanced computational machinery. Chapter 1 referred to the pioneering yet unsuccessful Argo Venture project, dating from a third of a century ago, that would have been staged in the physical world. Since then, the old tradition of enacting role-playing games in our real lives has merged with online communications, for example in the very popular virtual scavenger hunt conducted across real geography, *Pokémon Go*.[32] Called *LARPs*, for *live action role-playing*, this tradition has spawned *pervasive LARPs* and *multi-modal LARPs*, now evolving into *augmented reality simulations*.[33] Thus we can imagine a possible Argo Adventure, in which people experienced Earth as if it were Mars, using Internet-connected mobile devices to transform the environment, perceptually or conceptually at least, to render it realistically Martian.

A decade before Argo Venture, the last human trip to the Moon took place, and politicians have ever since claimed that we are on the way to Mars, and yet we never seem to get there. Chapter 4 mentioned the presidential ceremony attended by Buzz Aldrin, in June 2017, at which the latest US administration set the same old goals again, and by the end of that year, the 2010 Space Policy Directive that had intended the government to "set far-reaching exploration milestones" was updated to include these: "Lead an innovative and sustainable program of exploration with commercial and international partners to enable human expansion across the solar system and to bring back to Earth new knowledge and opportunities. Beginning with missions beyond low-Earth orbit, the United States will lead the return of humans to the Moon for long-term exploration and utilization, followed by human missions

[32]https://en.wikipedia.org/wiki/Pok%C3%A9mon_Go, accessed December 2017.

[33]Walther (2005), Jonsson et al. (2006); https://en.wikipedia.org/wiki/Live_action_role-playing_game, accessed December 2017.

to Mars and other destinations."[34] Each step in that set of milestones could be the basis of a suite of social simulations, whether fully computational or also including human participants. Yet after all these many years of delay, perhaps our efforts should be devoted to simulations of the political process, to explore what changes in governmental institutions would be required to commit humanity to a truly cosmic quest.

References

Bainbridge, William Sims. 2011. *The Virtual Future,* 88. London: Springer.

Bainbridge, William Sims. 2015. *The Meaning and Value of Spaceflight,* 198–201. Berlin: Springer.

Burn, Andrew, and Diane Carr. 2003. Signs From a Strange Planet: Role Play and Social Performance. In *Anarchy Online in Proceedings of COSIGN-2003,* 14–23.

Castronova, Edward. 2005. *Synthetic Worlds: The Business and Culture of Online Games*. Chicago, Illinois: University of Chicago Press.

Carvalho, Vincius Marino. 2015. Leaving Earth, Preserving History: Uses of the Future in the Mass Effect Series. *Games and Culture* 10(2):127–147.

Chirot, Daniel. 1977. *Social Change in the Twentieth Century*. New York: Harcourt Brace Jovanovich.

Gibbon, Edward. 1896. *The History of the Decline and Fall of the Roman Empire*. London: Methuin, [1776–1788]).

Hobsbawm, Eric J. 1959. *Primitive Rebels: Studies in Archaic Forms of Social Movements in the 19th and 20th Centuries*. Manchester, England: Manchester University Press.

Hough, Jason M., and K. C. Alexander. 2017. *Mass Effect Andromeda: Nexus Uprising*. London: Titan.

Huizinga, Johan, and Homo Ludens. 1955. *A Study of the Play-Element in Culture*. Boston, Massachusetts: The Beacon Press.

Jonsson, Staffan, Markus Montola, Annika Waern, and Martin Ericsson. 2006. Prosopopeia: Experiences From a Pervasive LARP. in *Proceedings of the 2006 ACM SIGCHI International Conference on Advances in Computer Entertainment Technology*. New York: ACM. en.wikipedia.org/wiki/Live_action_role-playing_game, accessed December 2017.

Patterson, Christopher B. 2015. Role-Playing the Multiculturalist Umpire: Loyalty and War in Bioware's Mass Effect Series. *Games and Culture* 10 (3): 207–228.

Spengler, Oswald. 1926. *The Decline of the West*. New York: Knopf.

Sorokin, Pitirim A. 1937. *Social and Cultural Dynamics*. New York: American Book Company.

Toynbee, Arnold, 1947–1957. *A Study of History*. New York: Oxford University Press.

Thomis, Malcolm I. 1972. *The Luddites: Machine-Breaking in Regency England*. New York: Schocken.

Walther, Bo Kampmann. 2005. Atomic Actions—Molecular Experience: Theory of Pervasive Gaming. *Computers in Entertainment* 3(3)

Wolf, Eric R. 1969. *Peasant Wars of the Twentieth Century*. New York: Harper and Row.

Zakowski, Samuel. 2014. Time and Temporality in the Mass Effect Series: A Naratological Approach. *Games and Culture* 9(1):58–79

[34] www.gpo.gov/fdsys/pkg/FR-2017-12-14/pdf/2017-27160.pdf, accessed December 2017.

Index

W. S. Bainbridge, *Computer Simulations of Space Societies*, Space and Society,
https://doi.org/10.1007/978-3-319-90560-0

Printed in the United States
By Bookmasters